China Internet Development Report 2017

Chinese Academy of Cyberspace Studies
Editor

China Internet Development Report 2017

Translated by Peng Ping

中国工信出版集团 電子工業出版社
PUBLISHING HOUSE OF ELECTRONICS INDUSTRY
http://www.phei.com.cn

Springer

Editor
Chinese Academy of Cyberspace Studies
Beijing, China

ISBN 978-3-662-57520-8 ISBN 978-3-662-57521-5 (eBook)
https://doi.org/10.1007/978-3-662-57521-5

Jointly published with Publishing House of Electronics Industry, Beijing, China

The print edition is not for sale in China Mainland. Customers from China Mainland please order the print book from: Publishing House of Electronics Industry

Library of Congress Control Number: 2018949052

This Springer imprint is published by the registered company Springer-Verlag GmbH, DE part of Springer Nature
The registered company address is: Heidelberger Platz 3, 14197 Berlin, Germany

Foreword

Today, a new technical and industrial revolution represented by information technology is emerging, contributing to the development of economy and society. Thanks to this historic opportunity, China's Internet has entered the fast lane of development. Especially, after 23 years of rapid development of its Internet, the country has worked out a road of Internet development and governance with Chinese characteristics and has become a big country toward Internet development and application attracting the world's attention. It has created its experience in Internet development and governance and has contributed to the world Internet development.

Against this backdrop, we have launched the compilation of *China Internet Development Report 2017* (referred to as "the *Report*" hereinafter), in which the compiler tries to study and analyze in detail China's Internet development by summarizing what the country has done, analyzing the status quo, and looking into the future, so that the *Report* will be the best choice for people to know about the past, present, and future of China's Internet. To this end, the compiler tries to:

Take General Secretary Xi Jinping's strategic thoughts on developing and strengthening China through Internet development as the theoretical basis and guideline for the compilation

Since the 18th National Congress of the Communist Party of China (CPC), General Secretary Xi Jinping, a visionary man, has put forward a series of new ideas, thoughts, and strategies based on China's practice of Internet development and governance, systematically illustrating the important theoretical and practical issues concerning cyber security and informatization and thus forming rich, profound, scientific, and systematic strategic thoughts on developing and strengthening China through Internet development. All these ideas, thoughts, and strategies are an important part of Xi's Thought on Socialism with Chinese Characteristics for a New Era. In particular, the 19th National Congress of the CPC has proposed building China's strength in cyberspace, and making China into a digital country and a smart society, promoting the integration of Internet, big data, and AI with the real economy, developing digital economy and sharing economy, and cultivating new growth points and new driving forces, which provides a guidance and guideline for the development of the national cyberspace cause in the new era. We take Xi Jinping's strategic thoughts on developing and strengthening China through Internet development as the spirit and main thread running through the *Report* to interpret Xi's ideas, philosophy, and thoughts so that the readers can accurately grasp the significance of his thoughts.

Take the practice of China's Internet development as the research and reality basis

Guided by Xi Jinping's Thought on Socialism with Chinese Characteristics for a New Era, especially his strategic thoughts on developing and strengthening China through Internet development, the country has achieved a lot in Internet development and governance. The cyberspace is becoming clearer day by day, the national cyber security defense is being consolidated, informatization's role in boosting the economic and social development is becoming prominent, China's power of discourse and influence in cyberspace keeps increasing and the Chinese people are having more sense of gain in sharing the benefits of Internet development. All these vivid achievements provide broad space and rich resources for the compilation of the *Report*. While illustrating the status quo of China's Internet development, and new practice and achievements in particular, the compiler tries to make an overall review of the country's Internet development in the past 23 years, especially in the years since the 18th National Congress of the CPC; therefore, the *Report* is a systematic manifestation of China's experience in Internet governance.

Take comprehensiveness, accuracy, and objectivity as the aim and principle for the compilation

In the process of compilation, we always follow the principle of making everything internationalized, authoritative, accurate, theoretical, and generalized, to show the Chinese academic circle's understanding of and thought on the Internet development. We try to make the Report comprehensive by doing an all-field and panoramic research on China's network infrastructure, information technology, digital economy, e-governance, cyber security, and international cyberspace governance; we try to make all the data accurate, authoritative, and up to date by collecting data from governmental agencies, industries, and research institutes; we try to make our assessment objective by setting up China's Internet Development Index on the basis of the mature index systems from both home and abroad, making assessment on cyber security and informatization in 31 provinces (autonomous region and municipalities directly under the Central Government) from six dimensions to comprehensively and accurately reflect the Internet development level throughout the country.

It is our sincere hope that the *Report* will provide a new drive for China's Internet development, and will be used as a new window for the world to know about China's Internet development, and a new reference about China's Internet development and the world Internet development.

Beijing, China
December 2017

Chinese Academy of Cyberspace Studies

Contents

Overview

Looking at the history of world civilizations, humanity has progressed along with agricultural, industrial, and information revolutions. Each of these industrial and technological revolutions has had a great and profound impact on our way of production and life. Today, information technologies represented by the Internet are experiencing rapid changes with each passing day. They have brought about new ways of social production, created new space for people's life, opened new horizons of State governance, and enhanced our ability to understand and shape the world. China is going through a historic process of rapid application of information technologies, attaching great importance to Internet development. Since connected to World Wide Web 23 years ago, it has, in keeping with the principles of proactive utilization, rational development, law-based regulation and assurance of security, strengthened IT infrastructure, developed digital economy, and made life better for its people through IT application.

Since the 18th National Congress of the Communist Party of China (CPC), the CPC Central Leadership, with Xi Jinping as the core, holding the right banner, has been making the right strategies, conquering one difficulty after another, forging ahead, and launching reforms and innovations. They try to make everything better by considering the overall domestic and international situation. They have been planning and promoting the general layout of "Five in One" (*economic, political, cultural, social, and ecological progress*), coordinating and promoting the strategic layout of "Four Comprehensives" strategy (*to comprehensively deepen reform, comprehensively advance rule of law, and comprehensively strictly govern the Party so as to provide strong impetus and guarantee for comprehensively building a moderately prosperous society*), carrying out new development ideas, further reforming difficult areas, and promoting ethical party and government construction and combating corruption. Thus, many problems have been solved that remained unsolved before, and many desirable things have been done that remained undone before, leading to historic changes in the CPC and State governance, and to the establishment of the guiding position of Xi Jinping Thought on Socialism with Chinese Characteristics for a New Era. The CPC Central Leadership stresses cyber security and information security, listing the perfection of the cyberspace leadership mechanism as one of the 60 reform tasks defined at the Third Plenum of the 18th CPC Central Committee. Thus, the Central Leading Group for Cyber Affairs was formed, who integrates and coordinates cyber security and informatization issues in politics, economy, culture, society and military affairs. CPC General Secretary Xi Jinping, a visionary man, has put forward a series of new ideas, thoughts, and strategies based on China's practice of Internet governance, systematically illustrating the important theoretical and practical issues concerning cyber security and informatization and thus forming rich, profound, scientific, and systematic strategic thoughts on developing and strengthening China through Internet development. On the 19th CPC National Congress, the "Two Step Development Strategy" was put forward, ushering in a new era of building socialism with Chinese characteristics. The Congress has made strategic deployments on promoting development of Internet, big data, AI, and sharing economy so as to turn China into an influential country in cyberspace and a leading one in digital development and smart society, which will guide China's Internet development in the years to come.

Achievements of China's Internet Development

Guided by Xi Jinping Thought on Socialism with Chinese Characteristics for a New Era, and his strategic thoughts on developing and strengthening China through Internet development in particular, the top-level design and general framework for the country's Internet development and governance have been basically finished, positive energy and main theme spread have become the trend in the cyberspace, which is becoming cleaner day by day. China's cyber security has been consolidated. In the country, informatization plays an outstanding role in driving and leading the economic development. Meanwhile, China's power of discourse and influence concerning cyberspace in the world have been strengthened, and the Chinese people have got more sense of gain in sharing the achievements of Internet development. All in all, we have achieved a lot in Internet development.

Network infrastructure has become a new type of public infrastructure facilitating economic and social development

General Secretary Xi Jinping points out that we should strengthen network infrastructure construction and in-depth information resource integration, and open up the "main artery" of information for economic and social development. Aware of the trend of the global Internet development, China is now implementing the "Broadband China" Strategy and launching the action of "Facilitating Faster and More Affordable Internet Connection". We are constantly promoting network infrastructure construction and upgrading, and improving the capacity of broadband access, Internet service quality, and application level. We are the leader in the world in broadband network scale and construction quality. By June 2017, the bandwidth of China's backbone network had exceeded 100 Tbps, and that of international export had reached 7,974.779 Gbps; the number of the Internet broadband access ports had reached 739 million; that of fixed broadband users, 322 million; and that of mobile communication, 1.36 billion, including 890 million 4G users.[1] Network infrastructure in rural areas has witnessed leapfrog development, general service in telecommunication has improved greatly, and all townships and 93.5 percent of administrative villages have access to broadband. Application infrastructure is developing fast, the node number of Content Distribution Network (CDN) has reached over 2400, and peak broadband reserve has exceeded 10 TB. The quantity of key Internet resource has seen great growth, with the number of IPv4 addresses amounting to 338 million, that of IPv6 to 21,283 (in terms of blocks)/32 (in terms of addresses), both ranking second in the world. The number of registered domain names (.cn) is 20.61 million,[2] ranking first in the world.

Breakthroughs have been made in Internet information technology

General Secretary Xi Jinping points out that China should speed up its self-dependent innovation of core technologies in information area, and thus to narrow the gap between China and developed countries and even to become the leader of the world in some IT areas. According to Xi, by implementing the innovation-driven development strategy, China should launch a series of key projects with long-term benefits to the overall development of the country, carry out with unified strength researches into key generic technologies of the Internet and into cutting-edge technologies, trying to make breakthroughs in high performance computing, quantum communication, artificial intelligence (AI), cloud computing, big data, kernel processor chip, operating system, and 5G technology. By the end of 2016, the country had had 227,365 valid invention patents in computer, communication, and other electronic equipment manufacturing, including 118,725 in the single year of 2016, putting China on the top of the world in terms of international patent registration four times in recent 5 years. Mozi Satellite, China's first quantum science experimental satellite, has been successfully launched; Beijing–Shanghai Trunk, the world's first quantum confidential communication network, has been opened, covering over 2000 kilometers; and Beidou Navigation System (BDS), a global satellite navigation system researched and developed by China itself, has been applied in a large scale, with two Beidou No. 3 satellites launched at the first time, leading the country into a new era of globally networking. China has taken an active part in the formulation of international telecommunication technology standards, playing a leading role in completing a number of IETF standards, making key breakthroughs in IPv4/IPv6 interconnection, cyber security, and routing protocols, and producing increasing influence on the next-generation core technologies of the Internet.

[1]Source: *Economic Performance of Telecommunications* January to June, 2017, Ministry of Industry and Information Technology of the People's Republic of China (MIIT); *40th China Statistical Report on Internet Development*, China Internet Network Information Center (CNNIC).

[2]Source: *40th China Statistical Report on Internet Development*, CNNIC.

New economy represented by digital economy is thriving

According to General Secretary Xi Jinping, cyber and information technology sector represents the new productivity and new development direction, so it is expected to take one step ahead in practicing the new development philosophy. China gives full play to informatization's role in driving and leading the development of economy and society and develops its digital economy, having made remarkable achievements in e-commerce, information service, network terminal equipment, Internet finance, sharing economy and Internet+ integration. In 2016, the country's digital economic gross scale reached RMB 22.58 trillion *yuan*, leaping to the second place of the world, and accounting for 30.3 percent of national GDP.[3] Digital economy has become the new engine and new highlight of China's economic growth. E-commerce, the most active sphere in the country's digital economy today, involved 467 million online shoppers in 2016, with its transaction volume amounting to RMB 26.1 trillion *yuan*[4] and online retail sales volume to RMB 5.2 trillion *yuan*, ranking first in the world. Sharing economy is booming, with the market turnover amounting to RMB 3.45 trillion *yuan*[5] in 2016. China's Internet finance is also developing rapidly, with the number of its electronic payment services amounting to 139.56 billion in 2016, and thus a service value of RMB 2,494.5 trillion *yuan*.[6] The country encourages, supports, and regulates the development of digital economy through political guidance and administration under laws to create a favorable market environment for information network business, which has led to the increase of international influence of a number of companies such as Alibaba, Tencent, Baidu, and Huawei, and the emergence of many new Internet companies with innovation energy. As of October 2017, 730 Chinese companies in this area have gone public at home or abroad, with a total market value exceeding RMB 17.5 trillion *yuan*. Among the world's top 20 Internet companies in terms of market value, there are 8 from China.

Development of e-government has helped to innovate social governance

General Secretary Xi Jinping stresses that we should speed up our e-government and encourage the governments of different levels to break the information barrier and improve their service efficiency so that the common people will make less effort to get more information even from afar and so that we can make things easier, faster and simpler. China has listed "Internet + e-government" into the 12 priorities under the *National Informatization Planning for the Thirteenth-five-year Plan Period*, having formulated policies such as *Guidelines on Promoting the Coordinated*

---

[3]Source: *White Paper of China's Digital Economy Development 2017*, Chinese Academy of Information and Communications Technology (CAICT).

[4]Source: *Report on China's E-commerce 2016*, Ministry of Commerce of the People's Republic of China.

[5]Source: *Report on China's Sharing Economy Development 2017*, Research Branch of Sharing Economy, State Information Center of China (SIC) and Sharing Economy Working Committee, Internet Society of China (ISC).

[6]Source: *Overview of Payment System Operation*, The People's Bank of China (PBOC).

Development of E-government and *Guidelines on Speeding up the Internet + E-government*. All the local governments are expected, in accordance with the planning made by the Central Government, to keep deepening the application of e-government, establish the mechanism for governmental interlink, information sharing, and business interconnection, promote the construction of online service platforms for public service integration, and provide the full-flow, one-stop online service across sectors and areas. By the end of 2016, out of 31 provinces, autonomous regions and municipalities directly under the Central Government, and Xinjiang Production and Construction Corps, 27 had established the provincial e-government service platform.[7] New media of e-government are witnessing rapid development. By June 2017, there had been 170,000 e-government Weibo accounts verified by Sina Weibo Open Platform, and 514,000 e-government WeChat public accounts. They are new means of enhancing information openness, interaction between the governments and the people, and supervision from the people, guaranteeing the citizens' rights to know, to participate, to express their opinions, and to supervise the governments.

Capacity of cyber security assurance keeps improving

General Secretary Xi Jinping points out that without cyber security there would be no national security and without informatization there would be no modernization. According to him, security ensures development, and development is what security is aimed at. China has formulated a series of laws, regulations, strategies, and plans including the *Cyber Security Law of the People's Republic of China* and the *National Cyber Security Strategy*, established and strengthened the cyber security accountability, cyber security review mechanism, and cyber security monitoring, warning, and response mechanism, amended the *National Emergency Response Plan for Internet Security Incidents,* and enhanced cyber security awareness and capacity of incident analyzing and attributing as well as of quick recovery after cyber attacks. The country has reinforced security protection of critical information infrastructure, and has, for the first time, launched nationwide inspection on the critical information infrastructure in the key areas like finance, energy, electric power, communication, and transportation to promote improvement and construction. It is speeding up standardization of cyber security. By the end of 2016, the National Technical Committee for Information Security Standardization (NTCISS) had issued 195 national information safety standards.[8] The cyber security industry scale keeps increasing, with over 1000 such businesses and RMB 135.5 billion yuan[9] of value in 2017. The first-level disciplines of cyber security have been established in universities. Seven demonstration projects in cyber security colleges have been launched and cyber security schools have been

---

[7]Source: *Blue Book of E-government: Report on China's E-government Development 2017*, Social Sciences Academic Press of China.

[8]Source: CCID Think Tank ("CCID" for short hereinafter).

[9]Source: *Blue Book of China's Cyber Security Development 2016-2017*, Chinese Academy of Electronics and Information Industry Development.

established in 21 universities. Insisting that the cyber security should serve and rely on the people, we attach importance to cyber security publicity and education, having held four sessions of "National Cyber Security Awareness Week", involving over two billion participants, and thus cyber security awareness and techniques have been dramatically improved nationwide.

Cyberspace is becoming clearer day by day

General Secretary Xi has pointed out that cyberspace is the common home of humankind. Making it better, cleaner, and safer is the common responsibility of the international community, according to the interests of the people. Positive energy spread in cyber content construction and administration is the general requirement, and the controllability is the absolute principle. We are enhancing positive publicity, innovating the dissemination means, managing the timeliness, appropriateness, and efficiency of public opinions in cyberspace, improving the transmissibility, leadership, influence, and credibility, and launching a large number of phenomenal news broadcasts and thematic publicity products conforming to the online dissemination features and the interest of Internet users. All of these measures have helped to increase the rate of arrival, reading, and thumbing up. We also pay attention to the cultivation and practice of the core socialist values. For example, we have launched the online dissemination of excellent Chinese culture and cultivate healthy online culture, so that the core socialist values can be adopted to increase the cohesion of society. We have adopted legal, administrative, and technical means to strengthen the cyberspace administration, improve the network communication law enforcement system and mechanism, promote Internet governance, opening and visiting in accordance with laws, and enhance the accountability of competent authorities for the local cyberspace administration and the responsibility of website entities, and advocate administration in a scientific and regulated way. Full play has been given to the roles of social organizations, businesses, and specialists, and the mode of network environment management involving different participants has been formed. The mechanism for linked tackling of illegal information and websites, and complaints from users has been improved. In 2016, 40 million complaints were tackled, and more than 92 percent of harmful information was disposed of. So far, 20 specific projects like "cleaning" and "seedling-protection" have been launched to dispose of harmful information, which has helped to make cyberspace cleaner.

China's power of discourse and influence in cyberspace keeps increasing

General Secretary Xi Jinping has illustrated the concept of the international cyberspace administration on many an occasion. He has proposed "four principles" for promoting the global Internet governance system reform and "five proposals" for constructing a community of shared future in cyberspace, which has contributed to global Internet development and governance. Guided by Xi's "four principles" and "five proposals", China has released *Strategy of International Cooperation in Cyberspace*. It enhances its communication and cooperation with other countries, and participates in and produces an influence on the progress of the international Internet governance. It has listed "digital economy" as the theme of G20 Summit for the first time in the history of G20 Summit, and released the *G20 Digital Economy Development and Cooperation Initiative*, in the formulation of which China was the coordinator. We have enhanced the almost totally self-reliant construction of the Internet governance platforms and opened and hosted three sessions of the World Internet Conference (Wuzhen Summit), with 4600 participants, including officials, specialists, entrepreneurs, and leaders from different countries and international organizations. Wuzhen Summit has become the international platform for interconnectivity between China and other countries and the Chinese platform for Internet sharing and shared governance. We have deepened the cyberspace cooperation in the framework of the UN, APEC, and BRICS, promoted the internationalization of Internet name and digital address distribution mechanisms, as well as bilateral communication and cooperation between China and the U.S., Russia, and Europe. Cooperation has been launched among countries along the Belt and Road, the construction of key projects like China-ASEAN Information Harbor, China-Arab States Online Silk Road Ningxia Pivot has been accelerated, and China's views on Internet governance have been made into common consensus. The country's power of discourse and influence in cyberspace has increased dramatically.

The Chinese people are sharing the benefits of Internet development

According to General Secretary Xi Jinping, the development of the Internet development must be oriented to benefit the people, who can afford to use the Internet and then have the chance to use it and use it well, thus having the sense of gain when enjoying the benefits of Internet development. By June 2017, the number of China's Internet users had reached 751 million and that of websites in the country, 506 million.[10] The Internet has become an indispensable part in our daily life, with hundreds of millions of Internet users obtaining and exchanging information on it. We have implemented as a whole five projects of poverty alleviation through the Internet, which has seen great success thanks to the targeted poverty alleviation and elimination. We are facilitating resource sharing and developing education through the Internet to realize education equality and improve the efficiency. In 2016, the online education market scale was RMB156 billion *yuan*. We are making information service convenient and beneficial to the people, who then can go through official procedures, do shopping, see the doctor, and pay on the Internet. By the end of 2016, over 400 cities had opened up online city service through WeChat and Alipay, covering 50 types of online service and over 400 million users.

[10] Source: *40th China Statistical Report on Internet Development*, CNNIC.

Assessment of the Internet Development in all Provinces, Autonomous Regions, and Municipalities Directly Under the Central Government

In this report, we set up China's Internet Development Index System covering infrastructure construction, innovation capacity, digital economic development, Internet application, cyber security, and cyberspace regulation. We have made an objective, authentic and accurate assessment of the effect and level of the Internet development in all provinces, autonomous regions, and municipalities directly under the Central Government. Scientific analysis and quantitative assessment have been adopted as the method. It is a comprehensive assessment index system in the Internet area, the first of its kind in China.

China's Internet Development Index is the specification, standardization, and indexation of General Secretary Xi Jinping's strategic thoughts on developing and strengthening the country through the Internet

Guided by Xi's thoughts, and based on the "six accelerations" (namely, to accelerate self-dependent innovation in network information technologies, to accelerate digital economy's drive for economic development, to accelerate the improvement of cyberspace administration, to accelerate cyber security protection, to accelerate the promotion of social governance through network information technologies, and to accelerate the enhancement of China's power of discourse and rule-making right in the world in terms of cyberspace) put forward by President Xi when he presided over the 36th Collective Study Session of the Political Bureau of the CPC Central Committee, our assessment covers the major content of cyber security and informatization. The Index System will make local governments more aware of their strategic goals and focus of Internet development, as well as their comparative, geographical and development advantages, so that they can take more practical and efficient measures to fulfill the local responsibilities, improve their local Internet development and governance, and promote the development of the national cyberspace administration in the direction of strengthening the country through the Internet.

China's Internet Development Index focuses on the domestic situation through a comparison with the international situation

Based on the domestic situation, by referring to the E-Government Development Index (EGDI), ICT Development Index, Global Cyberspace Security Index, and Networked Readiness Index (NRI) released by the UN, International Telecommunication Union (ITU) and World Economic Forum (WEF), and referring to the weight distribution and calculation method contained in the *World Internet Development Report 2017*, we have got the scores of the Internet Development Index of 31 provinces, autonomous regions, and municipalities in accordance with China's score (41.8), providing measurable and referable basis for all provinces, autonomous regions, and municipalities.

China's Internet Development Index is an objective, authentic and accurate assessment of the effect and level of the Internet development in all provinces, autonomous regions, and municipalities directly under the Central Government

To ensure the scientificity, accuracy, and authority of the Index (see Table 1), we have adopted all the data and materials provided by the local authorities and referred to statistics and materials provided by local departments, industries, research institutes, and enterprises. But, due to the limited data collection, our assessment in 2017 does not involve Hong Kong, Macao, and Taiwan. Besides, in view that the system has just been established, it is to be tested. We will adjust and improve the index system, enhance data collection, and expand the coverage, so that the assessment will reflect the reality more completely, accurately and scientifically.

Following the above weight distribution and calculation method, we have got the scores of the Internet Development Index of 31 provinces, autonomous regions, and municipalities of China (see Table 2).

Table 1 China's internet development Index system

Indicator	Key assessment factors	Weight %	Specification
Infrastructure construction	Number of broadband access ports and percentage of optical fiber users, number of 4G mobile base stations, number of wireless network accesses, and access speed	18	Local infrastructure construction level and Internet coverage of broadband, mobile and wireless network
Innovation capacity	Number of patents registered, accumulated patent efficiency, human resources, and investment in R&D.	16	Level , capacity and environment of local industrial innovation and local talents cultivation
Development of digital economy	ICT industry, income of telecommunication, e-commerce, Internet finance, and sharing economy	16	Development of local ICT industry and e-commerce
Internet application	Rate of individuals' use of the Internet for socializing, entertainment and finance, rate of corporations' use of the Internet, and online governments	25	Local individuals' use and corporations' use of the Internet, and e-government
Cyber security	Cyber security working mechanism, cyber security holes protection, phishing site intercept, and Internet users' cyberspace security awareness	13	Security and protection of local WiFi, websites, and personal information
Cyber regulation	Establishment of cyber organizations, number of governmental Weibo and WeChat accounts, and the number of staff engaged in cyber affairs	12	Capability of constructing and managing local cyber regulation organizations and personnel

Table 2 Ranking of the provinces, autonomous regions, and municipalities in the assessment of the Internet development

Ranking	Region	Index of infrastructure construction	Index of innovation capacity	Index of development of digital economy	Index of internet application	Index of cyber security	Index of cyberspace regulation	Total score
1	Guangdong	5.75	7.69	7.65	16.27	3.83	3.99	45.19
2	Beijing	6.66	5.68	5.51	18.93	3.57	4.35	44.70
3	Zhejiang	7.34	6.73	6.46	14.60	2.93	4.31	42.36
4	Jiangsu	7.44	7.60	6.62	13.03	3.57	3.65	41.92
5	Shanghai	6.47	4.18	4.50	16.81	3.05	3.93	38.94
6	Fujian	5.53	2.89	3.89	14.46	3.66	3.00	33.43
7	Sichuan	5.19	2.79	3.65	12.50	4.17	3.74	32.04
8	Shandong	5.98	4.69	3.72	11.24	2.81	3.01	31.45
9	Tianjin	5.49	3.50	2.67	12.40	3.91	2.72	30.69
10	Hubei	3.94	2.69	4.35	11.86	4.39	3.11	30.34
11	Anhui	2.52	3.20	3.26	11.98	5.22	2.68	28.86
12	Henan	6.49	2.94	2.82	8.95	3.27	3.73	28.20
13	Shaanxi	3.32	2.50	2.94	12.50	3.58	3.23	28.07
14	Hunan	5.39	2.50	2.50	12.43	2.70	2.55	28.06
15	Liaoning	6.36	2.42	3.22	10.24	2.56	3.02	27.81
16	Hainan	2.89	2.01	2.46	13.98	3.63	2.66	27.62
17	Chongqing	5.76	2.48	3.08	10.08	3.32	2.84	27.57
18	Shanxi	4.34	2.11	2.76	10.19	4.59	2.85	26.83
19	Hebei	6.55	2.31	2.34	8.47	3.62	3.02	26.31
20	Guizhou	4.89	2.63	2.26	9.86	3.49	2.87	25.99

(continued)

Table 2 (continued)

Ranking	Region	Index of infrastructure construction	Index of innovation capacity	Index of development of digital economy	Index of internet application	Index of cyber security	Index of cyberspace regulation	Total score
21	Qinghai	3.94	2.40	2.32	9.25	5.32	2.60	25.81
22	Guangxi	4.61	2.18	1.97	10.09	3.70	2.92	25.48
23	Xinjiang	3.63	2.78	2.44	8.21	4.51	3.84	25.41
24	Jiangxi	4.30	2.49	2.51	9.86	2.92	2.57	24.64
25	Inner Mongolia	4.39	1.61	2.29	8.90	4.31	2.89	24.41
26	Ningxia	5.20	1.87	2.64	7.49	4.48	2.69	24.37
27	Heilongjiang	3.94	2.18	1.99	9.27	2.93	2.81	23.12
28	Yunnan	1.90	2.12	2.60	10.56	4.33	1.24	22.74
29	Xizang (Tibet)	1.97	0.77	1.69	7.46	6.99	3.49	22.37
30	Jilin	4.73	1.85	1.83	8.35	2.76	2.77	22.30
31	Gansu	3.22	2.35	2.39	7.96	4.45	1.91	22.28

In general, since the 18th CPC National Conference, all the provinces, autonomous regions, and municipalities have been carrying out General Secretary Xi Jinping's strategic thoughts on developing and strengthening the country through the Internet. They are, in accordance with the unified arrangement by the CPC Central Committee, enhancing the cyber content construction, cyber security guarantee, and informatization through a series of new creations and practices based on their local reality, and have achieved great success.

(1) **In terms of infrastructure**
The coverage of information infrastructure and the number of Internet users have increased, and the gap between urban and rural areas in information infrastructure is narrowing since the local governments have increased the fund and political support. In general, there is a positive correlation between the information infrastructure development and the local economic and social development and investment level, and in some regions, the information infrastructure development is even ahead of the local economic and social development and investment level. For example, Guangdong, Sichuan, and Fujian take the lead in the FTTH (Fiber to The Home) and 4G network; Guizhou, Hubei, and Shanxi are constructing their data center and new-type infrastructure to lay a solid infrastructure foundation for the Internet development.

(2) **In terms of innovation capacity**
All regions participating in the assessment have begun to take network information technology innovation as the grip for the local innovative development. For example, Beijing and Shanghai, by taking the advantage of having more talents and technology, take the lead in developing AI technology, having made breakthroughs in AI chips; Inner Mongolia and Shaanxi have increased their support for innovation projects in the areas concerning information, encourage patent registration and application, and support research institutes and Internet enterprises in establishing core technology R&D centers and industrial alliances, and promote R&D design mode revolution; Guizhou, Shandong, and Jilin support high-end information industry development, and especially support telecommunication, chip, and software industry development with leading technology.

(3) **In terms of development of digital economy**
All regions participating in the assessment take e-commerce and sharing economy as the development focus, so e-commerce is increasing and seeing little regional difference. Regions leading in e-commerce attach importance to mode innovation, cultivate unique features, increase the scale, and go global. For example, Zhejiang takes the lead in digital economy scale by establishing experiment and advance zones for information economy development, and seven centers including the International E-commerce Center, National Intelligence Logistics Center, and National Digital Content Industry Center. Chongqing is trying to become one of the national pilot cities of the provincial

level in terms of e-commerce, and Henan is cultivating local e-commerce platforms, trying to take the lead in the country in terms of industrial segment of e-commerce.

(4) **In terms of Internet application**

Application dimension of new technologies represented by the Internet has been increasing, with diversified individual application, in-depth industrial application and efficient governmental application. For example, Beijing, Shanghai, Hunan, and Heilongjiang are fostering their new local media and trying to establish new models like Internet+ senior citizens caring and online teaching. The client use ratio of Toutiao, The Paper, and Inke keeps increasing. Enterprises in Zhejiang, Guangdong, and Shaanxi have launched intelligent production and network-based management by using big data, industrial cloud platforms, and industrial software. Inner Mongolia, Ningxia, and Henan are promoting precise production of agriculture and animal husbandry by using the prairie big data system and agricultural intelligent networking system. Guangdong has set up the assessment mechanism for governmental websites, and Guangxi and Hainan have set up governmental information sharing platforms and provide one-stop online government services, so their e-government has seen great success.

(5) **In terms of cyber security**

Since the cyber security accountability system was adopted, especially since the Critical Information Infrastructure Inspection and China Cyber Security Week were carried out, the cyber security awareness and guarantee strength have obviously increased in different regions, but there remain problems like insufficient organization guarantee and insufficient talents support. Provinces like Shaanxi has strengthened cyber security monitoring and found and blocked cyber attacks. Beijing, Guangdong, and Sichuan have seen rapid development of cyber security industry. Jiangsu and Hubei have established cyber security colleges to enhance the cultivation of talents in that field, with great potential in development.

(6) **In terms of cyberspace regulation**

All local governments are, in accordance with the unified requirement of the Central Government, promoting the establishment of provincial and prefectural cyberspace administration organizations, innovating the cyberspace administration patterns, strengthening team-building, and fulfilling local accountability. For example, Beijing, Shanghai, Guangdong, and Qinghai have constructed website classification and categorization administration systems, urging websites to fulfill their accountability and increasing the pertinence of the administration. Jiangxi has extended cyberspace administration organization construction to the county level, and channeled cyberspace administration into the local assessment indicator system. Tibet became the first to set up and improve three-level cyberspace administration organizations, namely, the autonomous region level, the prefectural level, and the county level, which has helped to form an integrated cyberspace public sentiment monitoring, guidance and governance platform system and thus has improved the cyberspace administration capacity. Xinjiang attaches importance to institutional construction and team-building, punishment and governance of illegal information spread. They have achieved great success in cyberspace crime punishment and piracy and patent infringement punishment. Ningxia, Chongqing, and Qinghai have, in accordance with the local needs for the Internet development and governance, launched a series of laws, and regulations on mobile Internet administration and live streaming service.

China's Experience in Internet Development

China attaches importance to both development and governance of the Internet. After 23 years of development, the country has made historic achievements, witnessed historic changes and found a path of Internet governance with Chinese characteristics, thus contributing China's experience and wisdom to the global Internet development and governance. The country's experience in Internet development can be summarized as follows:

Ensuring the CPC's leadership in cyberspace affairs

The leadership of the Communication Party of China (CPC) is the greatest advantage for the socialist system with Chinese characteristics and for the rapid development of China's Internet. Since the country's full-function access to the Internet in 1994, its Internet has developed from nothing to something, from small to big, and from weak to strong. During the process, the CPC is not only the decision-maker on the country's access to the Internet, but also the planner, organizer, promoter and navigator of its network construction, development, and administration. Since the 18th CPC National Congress, the CPC Central Leadership with Xi Jinping as the core has been attaching great importance to cyber security and informatization. They have taken the improvement of the cyberspace administration leadership mechanism as one of the 60 reforms defined at the Third Plenum Session of the 18th CPC National Congress. They founded the Central Leading Group for Cyberspace Affairs, of which Xi is the director and Li Keqiang and Liu Yunshan the deputy directors. General Secretary Xi Jinping has chaired a series of meetings and symposiums, such as the Meeting of the Central Leading Group for Cyberspace Affairs, the Symposium on Cyber Security and Informatization, and the 36th Collective Study Session of the Political Bureau of the CPC Central Committee, putting forward a series of new ideas, thoughts, and strategies about cyber security and informatization, drawing the great blueprint for China's Internet development, and thus defining the direction and guidelines for cyber security and informatization of all sectors.

Adhering to the people-centered development philosophy

The Internet has become the new space for people's study, work, and life, and the new platform for obtaining public services. Over the past 23 years, China has been pursuing the basic principle of developing the Internet to benefit its people. By developing the Internet, the country is speeding up the coverage of information service, reducing the application cost, carrying out poverty alleviation, and meeting the expectation and need of the people. It is enhancing content construction and governance of the Internet, punishing illegal online actions and crimes, and creating clean cyberspace for Internet users, especially young people. The Party is carrying out the mass line, using the Internet to gather the public opinions, unite the people, carry out some work, and mobilize the initiative and enthusiasm of the people, who contribute to and enjoy the Internet development results.

Taking innovation as the first drive for Internet development

Cyberspace administration represents new productivity and new development direction, so it must be ahead in new idea practice. In developing its Internet, China is carrying out Five Development Concepts, namely, innovation, coordination, greenness, openness, and sharing, with innovation always in the very first place. The country is brave enough to innovate and reform. It is speeding up the innovation in systems and mechanisms, ideas, technologies, culture, and application by making use of the features and advantages of the Internet. It releases new demand, creates new supply, cultivates new drive, optimizes the allocative efficiency of labor, capital, land, technology, and management. It also cultivates and attracts a large group of Internet entrepreneurs, talents, and leading figures in this field, and enhances the thriving of new technologies, new industries, and new ecology, so that the Internet has become the rising field with the most active innovation and the most rapid development.

Pursuing both opening up and self-reliance

To tackle the relationship between opening up and self-reliance is the basic experience of China over the past 30 years and the solid foundation for the rapid Internet development of the country. As is known, core technology is a nation's treasure, so China, with self-dependent innovation, self-reliance, and self-improvement, would like to cooperate with other countries in developing fundamental technologies and general technologies, asymmetric techniques, and trump-card techniques, core technologies like cutting-edge technology and game-changing technology, and thus we can make breakthroughs in R&D of chips for cell phones, supercomputers, servers, and searching technologies. Besides, China keeps constructing everything in opening up and introduces and absorbs advanced technologies and fruits from global network information areas, and makes its Internet development keep up with the world's pace and trend, so that the country has become the leader in many areas of the Internet.

Pursuing paralleled development of administration and service

China's Internet businesses have been developing from small to big, and from weak to strong, playing an important role in stabilizing the growth, promoting employment, and improving the people's life. To develop the Internet businesses healthily is not only the goal of the entrepreneurs, but also the requirement of the country's development. China attaches importance to the dominating role of businesses in the Internet development. We encourage, support, and regulate their development and guide and manage them in accordance with laws. We strengthen the construction of laws concerning the Internet, speed up the establishment and improvement of the law and regulation systems concerning the Internet, strengthen Internet governance in accordance with laws, and urge the businesses to fulfill their accountability. Besides, China has formulated a series of policies and measures to support the development of the Internet businesses. It speeds up the reform in examination and approval, funding and patent systems, breaks the barriers in systems and mechanisms, establishes the unified market, reduces the enterprises' burdens, and creates favorable environment for their development. Chinese Internet businesses attach equal importance to economic and social effects, volunteer to be supervised by the government, undertake social and moral responsibilities, reward society and serve the people, and form the joint force for the government–business cooperation and coordination and for the Internet development.

Ensuring both security and development

Cyber security and development are the two wings of one bird and two wheels of a cart. China always takes the proper tackling of the relationship between cyber security and development as the priority among priorities of the work concerning cyber security and informatization. The country insists on unified planning, layout, facilitation, and implementation, and ensures development through security and vice versa. We have formed the correct outlook on cyber security, which is always taken as a priority in informatization, and set up the cyber security accountability. We have carried out a thorough examination on the critical information infrastructure, strengthened self-dependent innovation, security review and opening and cooperation, and complemented the shortcomings in security with core technology breakthroughs, while consolidating security through critical information infrastructure protection to safeguard complete, dynamic, open, relative, and common cyber security, and thus to guarantee security for the rapid development of the Internet.

Trend of Future Development and Suggestions on Policies

From October 18 to 24, 2017, the 19th CPC National Congress was held. General Secretary Xi Jinping, on behalf of the CPC Central Committee, gave a report at the Congress. In the report, he analyzed the situation of the world and China today, put forward the Two Centenary Goals and the new "two-step" strategy, described the general blueprint of the socialist modernization, and made overall, strategic, and prospective plans, opening up a new era for the socialism with Chinese characteristics. The report also makes strategic planning for promoting the Internet, big data, AI, and digital economy as well as for constructing a strong Internet country, a digital country, and a smart society, providing fundamental principles and action guidelines for the country's Internet development. The coming 30 years will be a great strategic period of opportunity for China's Internet development. At the new historic starting point, the Internet development of China should be guided by Xi Jinping Thought on Socialism with Chinese Characteristics for a New Era, the spirit of the 19th CPC Congress and the people-centered approach. We will speed up the implementation of the strategy on developing and strengthening the country through the Internet, promote the construction of digital China, and ensure more complete and balanced, and fuller and better sustainable development of the country, so that the Internet can better benefit the country and its people, and thus it will contribute more to the modernization of socialism with Chinese characteristics.

The Internet development should be shifted from coverage acceleration to quality improvement, and the Internet should be a new facility and platform for economic and social development, with the construction of "digital China" as the leading part

The Internet in China, after 23 years of rapid development, is now experiencing the shift of changes of capacity to changes of quality. At present, the difference between provinces in the field of the Internet is decreasing, the increase in the number of users is not so dramatic, and the social investment in the Internet is becoming rational. There is limited sustainability in the rapid growth resulting from the bonus of facilities, population and capital. In the future, a "Digital China" should be built with high standards, quality, and level, the Internet should play a driving and leading role in deepening the supply-side reform, the network infrastructure should be completely upgraded, the promotion of the commercial layout of the new-generation information and telecommunication technology should be accelerated, the next-generation Internet construction should be sped up, and IPv6 should be laid out in an all-around way. We should make the network infrastructure fast, integrated, intelligent, and open, set up infrastructure platforms, data platforms, and resource platforms which can improve the economic development quality and efficiency to push the development of the primary, secondary, and tertiary industries into the top of the global value chain.

AI has become the new "blue sea" of the network information technology. We should occupy the commanding point by taking the R&D of critical and basic software and hardware as the focus

In the new active period of the global information technology development, cloud computing, big data, Internet of Things (IoT), AI, and other new-generation information technologies are developing fast. In particular, AI, as a cutting-edge comprehensive discipline integrating computer science, statistics, cerebral neuroscience, and social sciences, will be the leader of the next wave of the world's science and technology revolution as well as of the in-depth technical revolutions in all industries. It will bring about big leap in technology, data, and industries. Technically, China's high-performance computing capacity is increasing fast, its deep learning algorithm is evolving, and its mobile Internet is accelerating. The leap of these comprehensive technical capacities makes the breakthroughs in artificial technology possible, so we should concentrate our effort on the R&D of basic computing, chip/processor technology, intelligent operation system, computer vision and image technology, high-speed computing and large-scale data processing technology, and intelligent robots, which are all related to the development of AI technology, occupying the commanding point of the new-found information technology development. Industrially, China's mass user data will become the absolute advantage of AI development. Data computing, processing, and application of AI will create more high-value data for society. We should shift the investment focus from "Internet +" to "AI +", launch a new-found wave of innovations and new businesses and a number of phenomenal AI products meeting the requirement of multiple applications and intelligent service to boost innovations and revolutions in industrial organization pattern, service model, and business model of different industries.

Digital economy is developing fast, so we should make use of the trend of the new-round economic globalization to expand China's digital economy development space

China's digital economy development speed and level are in the leading position of the world, and such economy boasts the potential for sustainable growth. In the future, with the deep integration of the Internet, big data, AI and the real economy, a new-round economic globalization driven by data will accelerate, globalization of data trade, e-commerce, Internet finance, and online service will be unavoidable, cross-border platform trade volume, demand volume, and user number will be increasing by leaps and bounds, more small and medium-sized businesses and individuals will share the benefits of economic globalization, and platform economies will replace traditional multinationals and become the most influential international trade platforms. China should seize the opportunity to speed up the construction of the new-type open digital economy systems and mechanisms that combines advantages from both home and abroad and benefits the world. The country will promote intelligent agriculture, manufacturing, online service and Internet finance, and form new growth point and new drive in mid and high-end consumption, sharing economy, modern supply chain, and human capital service. It will lose no time in studying and formulating international rules and standards in terms of cross-border data trade, e-commerce, service trade, and Internet finance. It will also promote the establishment of the Global Digital Economy Development Union, guide Chinese Internet businesses in participating in the construction of the Belt and Road Initiative, free-trade ports, and multi-lateral free trade zones, and further expand the space for its digital economy development and thus improve the development level.

Paid content will become the new hot spot of the Internet media profit growth, and we should promote creative transfer and innovative development of Chinese culture

With the increase of cyberspace copyright awareness and the formation of the digital content consumption habit, paid online subscription will become more popular, involving news, video, literature and socializing, and other areas. It enables original content and quality content to be sold in a more convenient and efficient way, which, in turn, will lead to a bigger digital content production, service, and consumption market. "Content-priority" is becoming more important, so traditional media can make use of content production to find new opportunities for transformation and breakthrough and enhance the influence and yield rate of the content. New media can explore the areas with professional subdivision by optimizing user experience to occupy exclusive copyright resources and constantly increase the content "premium". Chinese culture, with a history of over 5000 years, combining revolutionary culture and advanced socialist culture, is becoming increasingly popular in the international community and among the people of the world, so we should take advantage of new network technologies and commercial patterns like qualified platforms, AI, and paid subscription to promote the creative transformation and innovative development and international communication of Chinese culture. We should tell the Chinese story properly to show an authentic, three-dimensional, and overall China and thus improve the country's soft power.

The construction of a smart society is closely related to the reform effect of the Party and State leadership system and mechanism, and we should promote data resources' opening up and utilization to enhance the modernization of the national governance system and capacity

With the development of the mobile Internet, the popularity of smartphones and the villagers' access to telecommunication service, the construction of a smart society will accelerate, and online service and life, and platform governance will become the new patterns of economic and social development, which will produce profound influence on the construction of politics, economy, culture, society, and ecology. In the future, e-government will be more intensive, and integrated e-government platforms throughout the country will become the mainstream. Therefore, it is necessary to promote the opening up and sharing of data resources, break data barricades between governmental agencies, between network platforms, and between governments and network platforms, and to realize cross-level, cross-region, cross-sector, and cross-system governmental administration and coordinated governance. We will keep deepening the development and unitization of governmental information resources and shift the static storage and simple application to dynamic management and deep analysis. We should find the hidden risks and hazards, fix the bug in supervision and improve the social governance effect by using big data and AI. The construction of a smart society will be expanded from individual pilot areas to the whole country. We will make social governance and public service more personalized, intelligent, and precise to meet the people's diversified demands and their longing for a better life.

Chapter 1
Comprehensive Improvement of China's Network Infrastructure

Chinese Academy of Cyberspace Studies

Network infrastructure is the "main artery" of the Internet, with great outward effects. Every technical upgrade and broadband expansion of Internet infrastructure will bring about leapfrog increase of network popularity and a new round of development of industrial Internet application. In recent years, China has been implementing the strategy of "Broadband China". Both the government and industries, through effort in a variety of fields, are trying to develop network infrastructure at a higher speed into intelligent infrastructure, with wide coverage and high popularity. Thus China has seen a great rise of its broadband access capacity and set up the world's largest 4G network; its key Internet resources and use efficiency have increased dramatically and Internet service quality, application level and industrial support capacity have caught up with the world's top level; the Initiative of Increasing the Speed while Decreasing the Expense has been launched throughout the country, the universal broadband telecommunication service compensation mechanism is being improved in rural and remote areas, and the digital divide between regions has been narrowed.

1.1 Network Infrastructure Has Become the Strategic National Public Infrastructure

1.1.1 Continuous Optimization and Evolution of Backbone Networks

1. **Top-level infrastructure of backbone networks keeps being optimized and evolved and interconnection quality keeps being improved.**

Since its full-function access to the Internet in 1994, China has been strengthening its backbone network construction, so that a variety of high-performance backbone

Chinese Academy of Cyberspace Studies
Beijing, China

Chinese Academy of Cyberspace Studies (ed.), *China Internet Development Report 2017*,
https://doi.org/10.1007/978-3-662-57521-5_1

networks have been formed covering all cities throughout the country, operated by three enterprises—China Telecom, China Unicom and China Mobile—and four nonprofit organizations—CERNet, CSTNet, CIETNet and CGWNet. Since 2000, top-level infrastructure for backbone network interconnection has been set up with three direct connection points, namely, Beijing, Shanghai and Guangzhou, as the main part, and with multiple exchange centers as the auxiliary. In 2013, seven other direct connection points for backbone networks were put into operation, namely, Chengdu, Wuhan, Xi'an, Shenyang, Nanjing, Chongqing and Zhengzhou. Network interconnection broadband keeps upgrading, network interconnection quality keeps improving, and network interconnection price keeps decreasing. Since the opening of the above-mentioned direct connection points, the number of internetwork routing hops has decreased by over 3, time delay by 60% to 80%, and packet loss ratio by 60% to 90%, and network traffic bypass in the middle and west regions has been improved and interconnection performance has increased by over 60%. Another three direct connection points, namely, Hangzhou, Guiyang and Fuzhou, were put into operation in 2016. Interconnection efficiency of backbone networks has been improved dramatically, and regional balance has been achieved, thus laying a solid foundation for the expansion of information service application.

2. **Capacity of backbone networks keeps increasing and they are being broadbandized and flattened.**

Relay bandwidth of China's backbone networks has witnessed the leap from Kpbs to Mbps and Gbps. Since 2000, with the upgrades of optical fiber technology, transmission and relay optical fiber of China's backbone networks has been extended to nearly one million kilometers, broadbandization of the backbone networks has been accelerated, with the highest single channel communication rate having increased from Kbps and 2.5G to 10 Gbps and 4G and even 100G, and the number of Internet backbone broad bands has seen the rapid growth of 50% to 100 % per year. The Internet structure keeps being optimized and backbone networks and Metropolitan Area Networks keeps being flattened, transformed from the star shape to the net shape. Network controllability has been strengthened and traffic efficiency and user service capacity of backbone networks have all been improved.

3. **Interconnection of networks has witnessed progress and international interconnectivity capacity keeps increasing.**

The bandwidth of international exit of China's Internet has seen rapid growth. By the end of 2014, the country had, through 22 inward and outward channels in the border cities, been connected by cable with 14 countries and regions, with its network now connected into Asia, Americas, Europe, Africa and Australia, and its overseas Point of Presence (POPs) totaling 72. Since the Belt & Road Initiative was proposed, China has strengthened its interconnection with the countries along the Belt & Road. It has promoted the cross-border optical fiber project construction between it and Burma, Nepal, Pakistan and Kyrgyzstan, and planned for the laying of new submarine cables to Asia, Europe and the United States and new international telecommunication inward and outward channels. China Unicom has set up

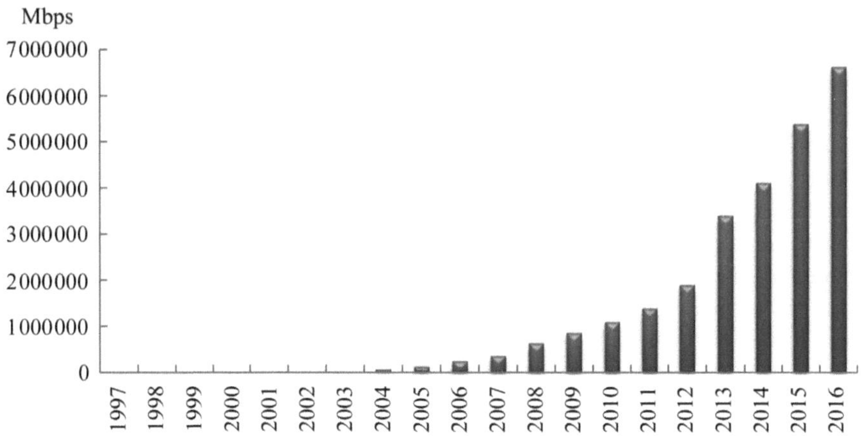

Fig. 1.1 China's international internet outward bandwidth from 1997 to 2016 (*Source* CNNIC)

44 POPs[1] in the hotspot areas along the Belt & Road, China Mobile has set up 29 POPs[2] along it, and China Telecom, 67 POPs,[3] having realized equivalent interconnection with major global telecommunication operators. From 2010 to 2015, China's international Internet inward and outward bandwidth witnessed an annual growth ratio of 15%, and in 2016, it witnessed such a growth of 23.1%, with 6,640.291 Gbps.[4] The country's international Internet outward bandwidth from 1997 to 2016 is shown in Fig. 1.1.

4. **The next-generation Internet construction has been carried out and has entered the starting stage of commercial use.**

China's next-generation Internet (NGI) development has seen four stages, namely, introduction, demonstrated application, trial commercial application and overall layout. IPv6 upgrading is being carried out. From 1998 to 2001, China successfully constructed its first IPv6 testbed CERNET-IP6v (1998), its first NGI Internet exchange center DRAGONTAP (2000) and its first test net NSFCNET (2001), realizing part of international NGI interconnection. In 2003, China's NGI Demonstration Project (CNGI) was launched, with its network covering 59 nodes of 22 cities and two international exchange centers, namely, Beijing and Shanghai. In August 2013, the project of National Next-generation Internet Demonstration Cities was launched, with over 20 key cities having comprehensively upgraded network

[1] *China Unicom: Four Measures to Promote the Construction of the Belt & Road*, http://finance.sina.com.cn/roll/2017-05-11/doc-ifyfecvz0909616.shtml.

[2] *China Mobile's Participation in the Construction of the Belt & Road*, http://news.163.com/17/0512/13/CK88823E00014AED.html#from = relevant.

[3] *Construction Chapter of China Telecom's Practice of the Belt & Road Initiative: Construction of "Information Silk Road through Global Layout*, http://www.sohu.com/a/139552103_354877.

[4] Source: CNNIC, *the 39th Statistical Report on China Internet Development.*

and service platforms and involved IPv6 broadband users. But generally speaking, these projects are not developing as fast as expected because of (1) inadequate support of China's Internet content providers (ICPs) for IPv6 and (2) insufficient IPv6 commercial network coverage and visits. In 2017, the country issued *Information Communication Industry Development Plan (2016–2020)* and *Information Communication Industry Development Plan—Internet of Things (2016–2020)*, which propose that by the end of the 13th Five-Year Plan period, IPv6 will have been able to be supported by major commercial websites, education and research websites and governmental websites, 80% of the users of top 100 Chinese mobile APPs will have been able to use IPv6, and the IPv6 traffic will have accounted for 5%. China's NGI has entered the stage of large-scale layout and application and will boost the new-generation information technology industry such as Internet of Things (IoT), Internet of Vehicles (IoV) and AI.

1.1.2 National Coverage of Broadband Network

1. **Technology of fixed broadband access keeps being upgraded and optical access has become the mainstream.**

China keeps increasing the investment in network construction. Since the dial-up access was opened to service in 1995, the country has witnessed fast evolution and upgrade of the fixed broadband access technology and the leap from low-speed broadband to high-speed optical fiber. In July 2012, the State Council issued the *National Plan for Strategic Emerging Industries during the 12th Five-year Plan Period*, which proposed that the "Broadband China" Project should be implemented, requiring that by 2015 urban and rural households should have the average broadband access speed of 20 M and 4 M respectively. In August 2013, the State Council released the *"Broadband China" Strategy and Its Implementation Plan*, which explicitly proposes that broadband network is the strategic infrastructure for China's economic and social development in a new era. Since then, copper cables have been rapidly replaced by optical fiber ones, optical access has become the mainstream technology and broadband network has entered a new stage. By June 2017, the number of Internet broadband access ports had reached 739 million and that of FTTH/0 ports, 599 million, from 75.6% of the total Internet broadband ports at the end of the previous year to 81%. Throughout the country, there are 322 million fixed Internet broadband users, among whom there are 261 million optical access users, accounting for 80.9%[5] of the fixed broadband users. China's fixed broadband users' number in comparison with that of the world from 2006 to June 2017 is shown in Fig. 1.2.

[5]Source: Ministry of Industry and Information Technology, *Economic Performance of Telecommunication Industry from January to June, 2017*, http://www.miit.gov.cn/n1146285/n1146352/n3054355/n3057511/n3057518/c5734845/content.html.

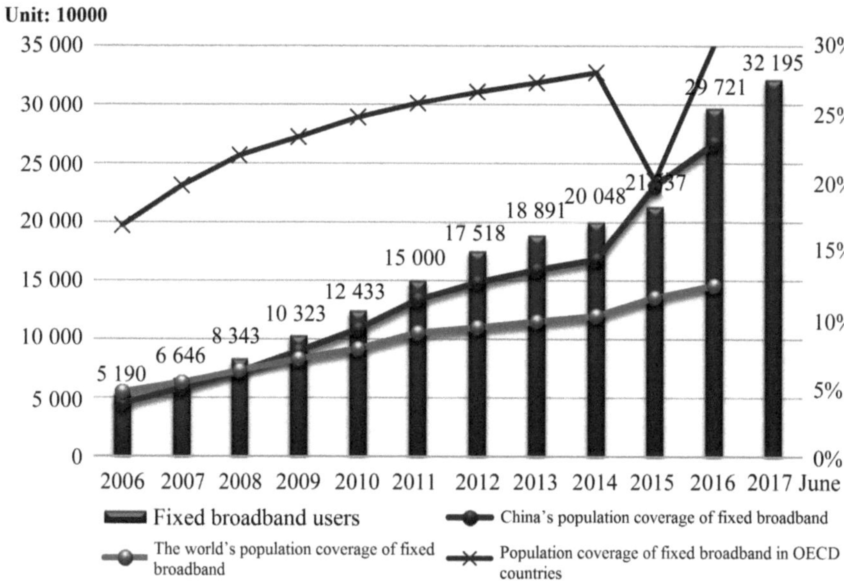

Fig. 1.2 Number of China's fixed broadband users in comparison with that of the world from 2006 to June 2017 (*Source* Ministry of Industry and Information Technology, the Word Bank and OECD)

Column 1 China's Three Telecommunication Operators

After some reforms in the field of telecommunication, today's telecommunication market of China is dominated by China Mobile, China Unicom and China Telecom.

China Mobile (with its full name as China Mobile Communications Group Co., Ltd.), founded on April 20th, 2000, is a mobile communication operator in the field of GSM, TD-SCDMA and TD-LTE mobile network service. It is a State-owned backbone enterprise founded on the basis of the divestment of mobile communication from China's telecommunication. It is on the top of the three telecommunication operators. Now, it fully owns China Mobile (Hong Kong) Company Limited. China Mobile Co., Ltd., a fully-owned subsidiary of China Mobile (Hong Kong), has fully-owned subsidiaries in 31 provinces (autonomous regions and municipalities directly under the Central Government) and Hong Kong Special Administrative Region and has been listed in Hong Kong and New York. By October 2017, its market value had totaled RMB1.6 trillion *yuan*.

China Unicom was founded in January 2009, a merged entity of the former China Netcom and China Unicom. It has branches in 31 provinces (autonomous regions and municipalities directly under the Central Government) and other regions and countries. Its business covers GSM, WCDMA and FDD-LTE mobile network service and fixed communication service, as well as network access service and other added-value telecommunication service. China Unicom is the only telecommunication operator who has been listed simultaneously in New York, Hong Kong and Shanghai. By November 2017, its market value had amounted to RMB 239 billion *yuan*.

China Telecom (with its full name as China Telecom Group Co., Ltd.), founded in September 2000, has its branches in 31 provinces (autonomous regions and municipalities directly under the Central Government) and Americas, Europe and China's Hong Kong and Macau, with its telecommunication service covering urban and rural areas of China and the world. It has established the world's largest CDMA 3G network with the widest coverage, the very first one for commercial use in the country. China's Telecom Co., Ltd. was listed in Hong Kong and New York in 2002. China's Communication Service Co., Ltd. was listed in Hong Kong in 2006. By November 2017, the market value of China Telecom had reached RMB 268 billion *yuan*.

2. **4G mobile communication network has seen basically full coverage.**

China's first GSM mobile telephone was put into operation in 1994, symbolizing the transfer of China's mobile communication technology from analog to 2G. In 2009, the country released 3G licenses and set up three 3G networks leading in the world's telecommunication development history. From 2010 to 2013, the number of 3G users in the country saw a dramatic growth, the compound annual growth rate of mobile Internet connection was over 50%, and 3G was replacing 2G. On January 18th, 2012, TD-LTE technology developed mainly by China was acknowledged by International Telecommunication Union (ITU) as one of the international standards for the fourth-generation mobile communication. In December 2013 and February 2015, 4G licenses were released. The world's largest 4G network was set up in China and 4G was replacing 2G/3G at a high speed. By June 2017, the total number of China's mobile subscribers had amounted to 1.36 billion, with 0.89 billion users of 4G, accounting for 65.1% of the total number of mobile subscribers.[6] The country's mobile Internet connection scale and its users' monthly traffic (2009–2016) are shown in Fig. 1.3. China's new mobile subscribers' profile (2009–2016) is shown in Fig. 1.4, and the number of China's mobile subscribers and new subscribers (2009–June 2017) is shown in Fig. 1.5.

3. **The 5G era is coming at a high speed.**

China attaches great importance to the development of 5G technology. It established in 2013 the 5G Promotion Unit (IMT-2020 Promotion Unit) made up of

---

[6]Source: Ministry of Industry and Information Technology, *Economic Performance of Telecommunication Industry from January to June, 2017*, http://www.miit.gov.cn/n1146285/n1146352/n3054355/n3057511/n3057518/c5734845/content.html.

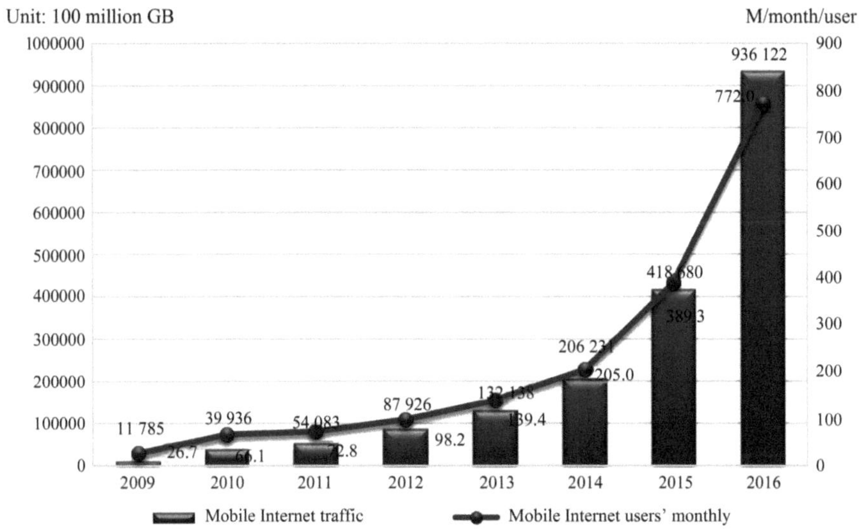

Fig. 1.3 China's mobile internet traffic scale and its users' monthly traffic (2009–2016) (*Source* Ministry of Industry and Information Technology)

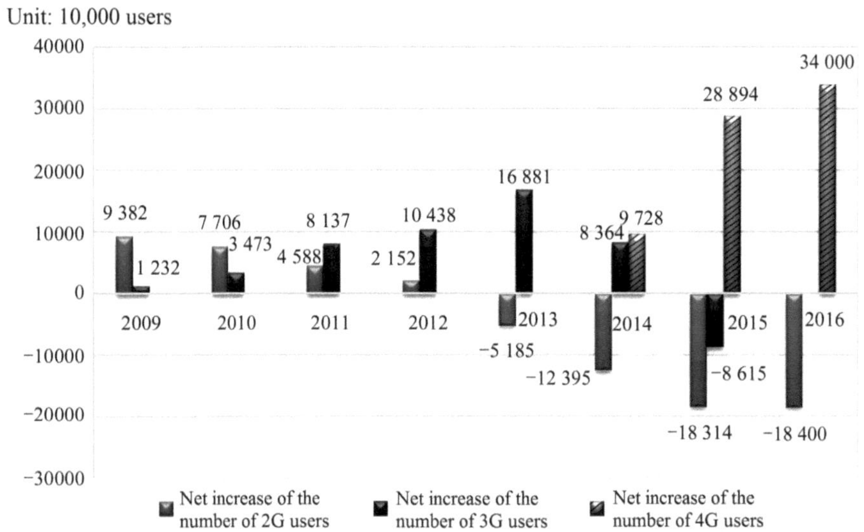

Fig. 1.4 China's new mobile subscribers' profile (2009–2016) (*Source* Ministry of Industry and Information Technology)

people from industries, universities and research institutes as well as some users so that the Unit can act as a platform to promote the R&D of 5G technology, test the technology plan and support the making of international standards. Since 2016, the Chinese Government has released the *13th Five-year Plan for Economic and Social*

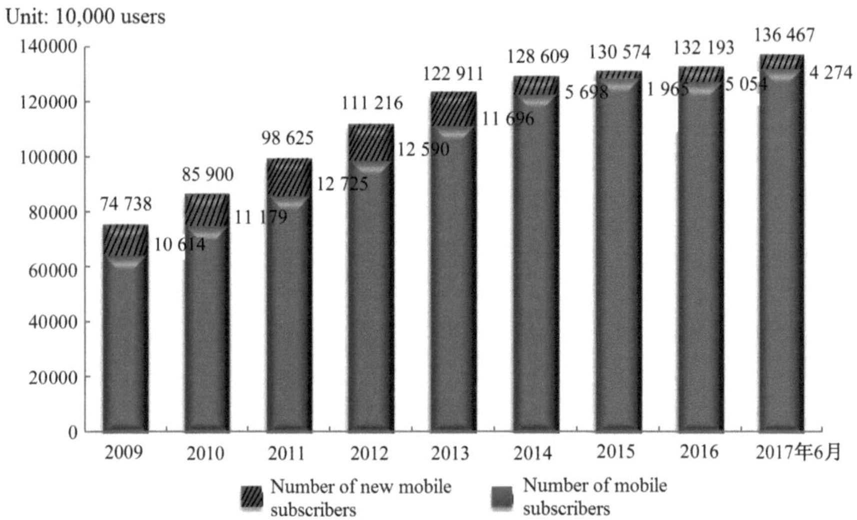

Fig. 1.5 Number of China's mobile subscribers and new subscribers (2009–June 2017) (*Source* Ministry of Industry and Information Technology)

Development of the People's Republic of China, *National Informatization Planning for the Thirteenth-five-year Plan Period*, and *Strategic Outline for the National Informatization*, all of which have made layout for the development of 5g technology. According to *Plan for the Development of Information Communication Industry (2016–2020)*, China will set up large-scale 5G testing networks in 2018, start the construction of 5G networks in 2019 and put the 5G networks into commercial use in 2020.

4. **China keeps implementing the Project of Increasing the Speed and Decreasing the Expenses.**

In May 2015, the Executive Meeting of the State Council launched the measures to speed up the construction of high-speed broadband networks to promote the increase of the speed and decrease of the expenses and facilitate the entrepreneurship and innovation, and improve the people's livelihood. The measures include acceleration of high-speed broadband network construction, optical access promotion and Broadband Village Project implementation, market opening and fair competition facilitation, and launch of pilot mobile communications resale and broadband access. From 2015 to 2016, China Mobile offered its users a benefit of over RMB5.6 billion *yuan*, the mobile phone data traffic charges of China Telecom decreased by 57.4% and the fixed broadband charges of China Unicom decreased by over 67%.[7] The year 2017 was the third year of China's implementing

[7]*The new-round increase of speed and decrease of expenses has been launched and here are the new plans of the three operators*, http://finance.sina.com.cn/chanjing/cyxw/2017-03-06/doc-ifycaasy7719405.shtml.

the Project of Increasing the Speed and Decreasing the Expenses. The three telecommunication operators have launched the policy of "rollover data service", keeps reducing their data traffic charges, and have reduced special line charges for small and medium-sized businesses and charges for major IDD calls. By June 2017, the proportion of over-20 M broadband users in the country had reached 86% and that of 4G users, 65.1%.[8] To be short, a series of achievements have been made in the Project of Increasing the Speed and Decreasing the Expenses.

5. **The universal telecommunication service capability has been improved.**

The Chinese Government always attaches importance to the development of information infrastructure in rural areas. In January 2004, Ministry of Industry and Information Technology launched *Universal Telecommunication Service in Rural Areas—Every Village Project Implementation Plan* ("Every Village Project" for short hereinafter), requiring that the telecommunication operators should divide up their work in different regions, raise fund themselves and make independent operation and maintenance to undertake communication network construction in rural areas. They have taken three three-steps strategies: the area coverage of network has been expanded from townships to administrative villages and then to natural villages; the service has been upgraded from telephone to Internet and then to broadband; the informatization has been extended from infrastructure construction to service platform construction and then to other services in rural areas. In 2014, the Broadband Village Project was jointly launched by National Development and Reform Commission, Ministry of Finance and Ministry of Industry and Information Technology, who, with funds raised through multiple channels, supported broadband construction and application in rural areas. By the end of 2014, broadband construction had been done in 3,000 townships and 150,000 administrative villages, Internet had covered 18,000 villages of special poverty and broadband had been installed in 6,400 schools and public interest organizations in rural areas. In October 2015, the Executive Meeting of the State Council launched the Compensation Mechanism for Reforming and Innovating Universal Telecommunication Service. Funds have been raised from the Central Government, local governments and enterprises, and the market mechanism has been adopted to support the broadband network construction in rural and remote areas. This is a great policy breakthrough in terms of the universal service mechanism of China, providing a strong support for the realization of the strategic goal of rural broadband development in 2020. By the third quarter of 2017, two batches of pilot projects of universal telecommunication service had been launched jointly by Ministry of Industry and Information Technology and Ministry of Finance. RMB 40 billion *yuan* had been invested in optical fiber installation in 130,000 administrative villages, among which 43,000 were registered poor villages. The implementation of the pilot universal telecommunication service has contributed to

[8]Fulfillment of major communication aims in June 2017, http://www.miit.gov.cn/n1146312/n1146904/n1648372/c5735472/content.html.

the rapid development of network in rural and remote areas of China, where the conditions have been greatly improved. 96% of administrative villages and 86% of poor villages have been covered by broadband.[9]

1.1.3 Application Infrastructure Promotes Internet Development

1. Content delivery network (CDN) coverage and the delivery capability are increasing.

Since 1998, CDN industry has experienced three stages, namely, trial period, steady development period and accelerated development period. In recent years, CDN has been developing at a dramatic speed and there have emerged a number of Internet businesses providing CDN service, major telecommunication operators have taken CDN as their value-added service, and CDN coverage and capability as well as service coverage keep increasing. By 2014, China had established over 2,400 CDN nodes of different kinds and there had been over 10 TB of peak broadband.[10] In December 2016, Ministry of Industry and Information Technology granted CDN licenses to Wangsu Science and Technology Co., Ltd. and Alibaba Cloud, marking that CDN service became licensed service. By June 2017, totally 16 businesses had obtained their CDN service licenses,[11] having thousands of nodes both at home and abroad, all of which had TB-level broadband scale. Alibaba, Baidu, Tencent, Wangsu, ChinaCache and Shanda all provide CDN-based cloud storage, cloud security, cloudwatch, cloud database, cloud host and cloud computing. CDN has become a key platform for CDN application of cloud service.

Column 2 CDN

CDN is the short form of Content Delivery Network. It is aimed to avoid any possible bottlenecks or factors on the Internet that may affect data transmission speed and stability to make the content transmission faster and more stable. Through a layer of intelligent virtual network based on the Internet constituted by node servers placed in different places of the network, the CDN system can re-transmit in real time the user's request to the nearest service node in accordance with network traffic and the connection between nodes and in accordance with loading and the distance to the user as well as

[9]http://www.miit.gov.cn/n1278117/n4310819/n4310832/c5882939/content.html.

[10]Source: *White Paper of CDN (2014)*, former Telecommunication Research Institute of Ministry of Industry and Information Technology.

[11]Ministry of Industry and Information Technology has issued the sixth batch of CDN licenses and the number of licensed businesses is now 16, http://www.sohu.com/a/164171420_488920.

the response time. It aims to help the user to get the content from the nearest source, make the network traffic smoother and improve the speed of the website's response to the user's visit.

2. **The scale of Internet data centers is increasing rapidly and cloud computing has become an important direction.**

With the development of China's Internet industry, data centers are developing from small simple storage and computing ones to Internet Data Centers (IDCs) that can support network information services. They have evolved from simple "data warehouses" into centers providing comprehensive services including data exchange, information processing and data analysis. Since 2004, all the basic telecommunication operators have increased their investment in data center construction, which is more large-scale, centralized and environment-friendly. The market scale keeps increasing by more than 10%. In January 2013, the *Guiding Opinions on Data Center Construction and Layout* was released. Major Internet businesses, basic telecommunication operators and some large enterprises in other industries are promoting the construction of cloud computing data centers, so the bearing and management capacity of cloud computing resources and services has been improved dramatically. From 2012 to 2014, China's data center market scale increased from RMB 8.04 billion *yuan* in 2011 to RMB15.6 billion *yuan* in 2014 and now there are over 400,000 such centers, among which there are approximately 1,000 computer rooms, 280,000 computer cabinets and over 3 million servers in business data centers.[12] By the end of October 2016, Ministry of Industry and Information Technology had issued 327 cross-provincial data center licenses and 844 provincial ones.[13]

1.2 The Quantity of Basic Internet Resources Is Increasing

1.2.1 China's IP Quantity Ranks Second in the World

1. **China's number of IP addresses increases rapidly but the effective use rate remains low.**

Since it obtained its first IPv4 address from the Asia-Pacific Network Information Centre (APNIC) in the 1990s, China has witnessed the fast growth of the number of its IP addresses. By June 2017, the country had got 338 million IPv4

---

[12]Source: CCID.

[13]*IDC Service Conference of 2016 was successfully held in Beijing,* https://www.csdn.net/article/a/2016-12-28/15842846.

addresses and 21,283 IPv6 blocks/32 bits,[14] ranking second in the world, but they cannot meet the need of development. First, the IP resources per capital remain low. With the increase of the number of Internet users, the network application level and Internet consumption capacity, people have higher and higher expectation of network service quality, which, in turn, will widen the gap between IP resources and Internet development. Secondly, all IPv4 addresses have been distributed but IPv6 has not been used in large-scale in China. According to APNIC data, by the end of 2016, there had been nearly 5 million IPv6 users (see Fig. 1.6) and the popularity of IPv6 had been only 0.7%, ranking 56th in the world, while Belgium enjoys the highest, 60% (Belgium), and then comes the United States, where the popularity is 34%.

1.2.2 The Number of Registered Domain Names Has Reached Tens of Thousands

1. **The number of registered domain names has reached 20 million, ranking first in the world.**

In May 1994, with the help of Professor Qian Tianbai and University of Karlsruhe of Germany, the Computer Network Information Center of Chinese Academy of Sciences finished the setting of the servers of the top domain name (. cn). After years of development, the number of registered domain names has grown drastically and infrastructure of domain name resolution keeps upgrading. In 2003,. cn was opened for registration and it witnessed rapid development soon. In July 2008, the number of registered domain names of.cn was 12.188 million, ranking first in the world. By the end of 2016, the total number of domain names in China had reached 42.28 million, with an annual increase of 36.3% (as shown in Fig. 1.7). Among them, the number of registered.cn domain names was 20.61 million, with an increase of 25.9%.[15]

2. **Chinese domain names are developing fast and a number of top Chinese domain names have been channeled into the global Internet domain system.**

In 1998, China Network Information Center (CNNIC) started the R&D of Chinese domain name technology, marking the beginning of integrating Chinese characters into Internet domain names. In 2010, the national top Chinese domain names 中国 and 中國 became root domain names and the global resolution of the two names was realized. Since new general top domain names were opened to application in 2012, a number of top Chinese domain names like.网络,.公司,.公益 and 政务 have been root domain names successfully and they have witnessed global resolution. In July

[14]Source: CNNIC, *40th Statistical Report on China Internet Development.*

[15]Source: CNNIC Statistical Report on China Internet Development, 2016. http://www.cac.gov.cn/2017-01/22/c_1120352022.htm.

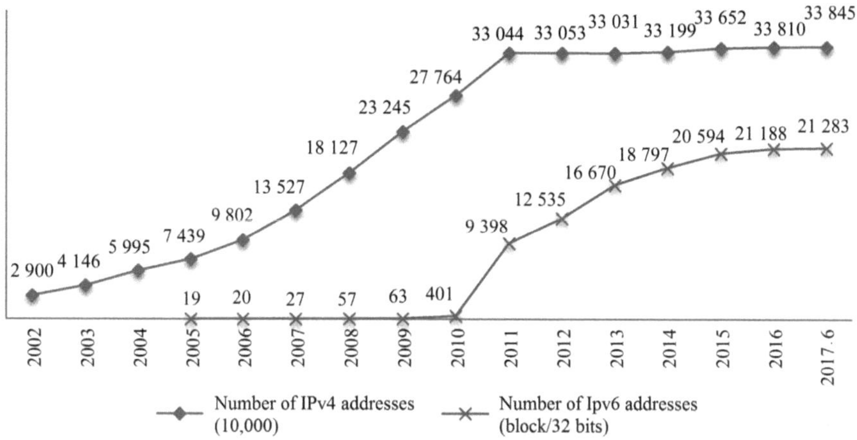

Fig. 1.6 Number of China's IPv4 and IPv6 addresses (2002—June 2017) (*Source* CNNIC, *Statistical Report on China Internet Development*)

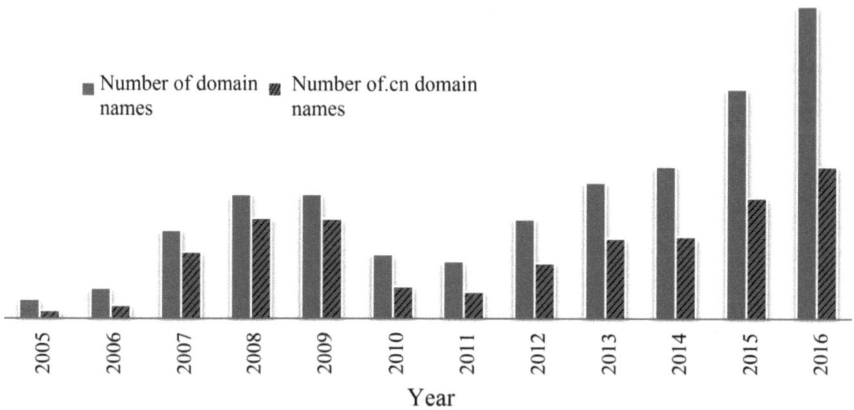

Fig. 1.7 Total number of China's domain names and .cn domain names (2002–2016) (*Source* CNNIC, *Statistical Report on China Internet Development*)

2014, new general top Chinese domain names like 公司 and 网络 was opened to registration. By December 2016, there had been 20.61 million.cn domain names, accounting for 48.7% of the total; and 474,000 中国 domain names, accounting for 1.1%.[16] The number of China's classified domain names is shown in Table 1.1.

[16]Source: CNNIC Statistical Report on China Internet Development, 2016. http://www.cac.gov.cn/2017-01/22/c_1120352022.htm.

Table 1.1 Number of China's classified domain names

Domain name	Number	Percentage (%)
CN	20,601,491	48.7
COM	14,345,243	33.9
NET	1,633,071	3.9
ORG	330,457	0.8
中国	474,115	1.1
BIZ	210,062	0.5
INFO	224,321	0.5
Others	4,456,942	10.5
Total	42,275,702	100.0

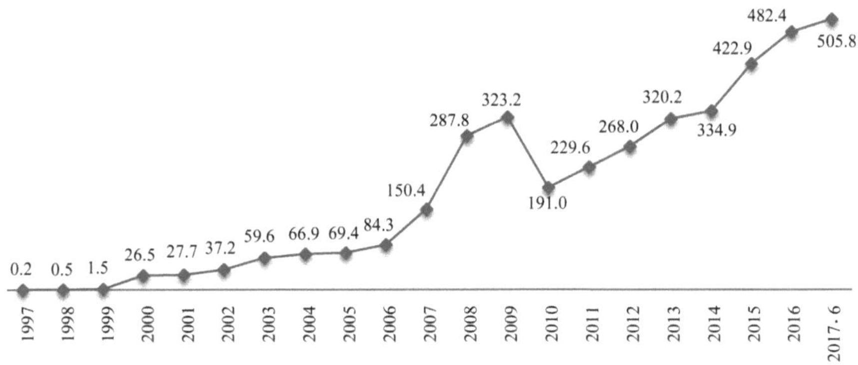

Fig. 1.8 Number of websites in China (1997–June 2017) (Unit: 10,000) (*Source* CNNIC, *Statistical Report on China Internet Development*)

3. **Domain name resources and excellent resolution services have been enriched, which has led to the establishment of a large number of websites.**

Since 1997, the number of China's websites has grown dramatically. Only in 2000, it increased from 15,000 to 265,000. With the development of the Internet industry, it rose to over three million in 2009. Since 2010, the Chinese Government has strengthened cyber security governance and the number of websites has decreased to 19.1 million, but the number of webpages and webpage bytes has been increasing and the quality of websites has been improved. By June 2017, the total number of websites in China had amounted to 5.06 million,[17] as is shown in Fig. 1.8.

[17]Source: CNNIC Statistical Report on China Internet Development, 2017. http://www.cac.gov.cn/2017-08/04/c_1121427728.htm.

Table 1.2 China's introduced root mirror distribution and government supervision

Root	Location	Year	Introducer	Management organization
F	Beijing	2003	China Telecom	ISC
F	Beijing	2003	CSTNET/CNNIC	ISC
I	Beijing	2005	CSTNET/CNNIC	Netnod
J	Beijing	2006	China Unicom (former China Netcom)	Verisign
L	Beijing	2012	CSTNET/CNNIC	ICANN

1.2.3 Resolution Service Keeps Being Upgraded to Support Website Service

1. **China keeps improving resolution service quality and has set up an effective domain name resolution service system.**

To fully connect China's Internet with the world, not only is the resolution of the domain name.cn needed, but also the resolution of other top domain names in the world is important. Since it got its first root server mirror in 2003, China has been introducing mirror nodes of root servers and those of top domains of.com/.net. By 2017, four mirror nodes of root servers of Internet domain names, F, I, J and L (as shown in Table 1.2) had been opened. Meanwhile, the country has set up domain name resolution service nodes in the world and has established resolution service facilities covering all domestic network operators and the Asia-Pacific Region, Europe and North America, which provide service for global users visiting China's websites and improve the users' experience. So far, the construction of comprehensive domain name resolution infrastructure has been completed.

Colum 3 Mirror Server

The mirror server has the same content as the primary server, but the former is placed in a different place to share the loading of the primary server. To put it simple, such a server is like a mirror. On the web, between the two servers or among more than two servers upgraded simultaneously with the same content, all of them are called mirror servers except the primary server.

Chapter 2
Breakthroughs Made in Information Technology

Chinese Academy of Cyberspace Studies

China always gives priority to technical innovation, implements innovation-driven development strategies, and gives full play to the role of all kinds of innovative bodies like Internet businesses. It gathers multiple forces to do research into key Internet technologies and has made great breakthroughs in many fields such as integrated circuits, basic software, core components, cloud computing, big data, IoT, NGI and satellite navigation. All these breakthroughs provide powerful support for China's developing from a big Internet country to an Internet power.

2.1 China's Communication Technology Development Is in Step with That of the World

After over twenty years of development, China has developed into a leader of the world in communication technology, including network equipment technology and international communication standard formulation. It has made great breakthroughs in quantum communication and space navigation technology. It is one of the countries with the highest number of patent applications, having made breakthroughs in the network equipment industry and mobile terminal manufacturing. A number of world-class communication technologies companies like Huawei and ZTE have emerged.

1. **China first caught up with the world, then made breakthroughs in terms of communication technology and later became the leader in some key areas.**

Since the key informatization projects and telecommunication network infrastructure construction projects were launched in 1994, China's network technology

Chinese Academy of Cyberspace Studies
Beijing, China

Chinese Academy of Cyberspace Studies (ed.), *China Internet Development Report 2017*,
https://doi.org/10.1007/978-3-662-57521-5_2

research and communication equipment research and manufacturing have been following the world's advanced level. We have made breakthroughs in key areas like optical communication and data communication, so we need not import everything as before. China's first set of 155 Mb/s SDH equipment and ATM switch were developed in 1995, so the country would no longer rely on foreign markets for such equipment. Optical communication has enjoyed rapid development and China has become a key exporter of communication technology, equipment and management systems. In 2006, 40 Gb/s SDH Optical Communication Equipment and System established by FiberHome was checked and accepted by the State and became the first optical communication equipment in the world conforming to ITU-T Standards. Huawei released carrier-class SDN controller SNC in 2013 and 32 wave-length 10G WDM-PON (Wavelength Division Multiplexing Passive Optical Network Prototype) to support the 320G total broadband of single fiber in 2016. In 2017, together with China Telecom, it accomplished the mid and long-range technology R&D and comprehensive testing of the first key high-capacity router technology (400GE). The equipment and technology mentioned above are the first in the world.

2. **China takes an active part in the formulation of international communication technology standards.**

With the rapid improvement of network technology innovation capacity and industrialization capacity, China is becoming more and more active in international standard formulation. Since it submitted the first IP international standard (ITU-T X.85) to ITU and gained approval in 2000, it has had increasing power of discourse in relevant fields. By November 2017, China had submitted more than 500 documents concerning standards to ITU-T.[1] In terms of IETF standards, the country finished seven Internet engineering tasks from 1996 (RFC 1922) to 2008 and led or participated in 185 standards in IETF from 2013 to October 2017.[2] Therefore, China had great potential in standard formulation. A number of departments, businesses and universities are now key members of international standard organizations like ITU-T and IETF and quite a few Chinese experts are now in the leading position in those organizations, which will contribute to the "going-out" of China's technologies and standards and to the overseas development of the country's network equipment suppliers and manufacturers.

3. **China's capacity of data communication technology innovation has improved.**

With the large-scale construction of network infrastructure and fast spread of Internet applications, China enjoys improving innovation capacity in the field of data communication technology. From 2012 to 2016, it ranks first for four times in terms of international patent application by communication equipment manufacturers. Chinese companies have successfully developed totally on their own

[1]Source: China's website of ITU, http://www.ituchina.cn/itut/zgwg/.

[2]IETF website, https://www.rfc-editor.org/rfc-index-100a.html.

high-end network processor (NP) chips and have been the first to release T-class routers in the world. Actually, at the beginning, they only imitated other products, but now they have surpassed many of their foreign counterparts and have begun to produce high-end products rather than low and mid-end ones. China's network equipment industry is developing fast and breakthroughs have been made in large scale. A number of fast-growing hi-tech businesses like Huawei and ZTE have emerged, whose technology of fixed broadband access systems and optical transmission systems takes the lead in the world. In recent two years, while the communication equipment industry of the world has been sluggish, Chinese businesses like Huawei and ZTE have maintained their momentum thanks to their innovation. In the first half of 2017, Huawei's sales revenue was RMB283.1 billion *yuan*, including 31.1 billion *yuan* of profits,[3] ranking first in the world; ZTE and FiberHome have also maintained their growth of more than 10%. Their scale has surpassed that of their overseas counterparts like Ericsson, ALE and Nokia.

4. **China's mobile communication technology is the leader in the world and its 5G technology R&D test has been comprehensively launched.**

China's mobile communication started from 2G, missing 1G, but breakthroughs were soon made in terms of 3G. It keeps up with the world in 4G technology and now it is working to be the leader in 5G. In 1990s, the country introduced from overseas 2G technical standards like GSM and CDMA and then succeeded in independent R&D of equipment and products of those fields. In May 2000, its TD-SCDMA technology as its own intellectual property was approved by ITU as one of the 3G standards and was launched into use for commercial purpose in 2009. As for the 4G development, China, by adopting the strategy of integrating technical innovation with standardization, developed TD-LTE standard, which is an integration of core TD technology and international mainstream standard LTE. In September 2010, such an integrated technology became the international 4G standard and an industry in that field was formed, made up of 10 system businesses, 14 chip businesses and over 80 terminal businesses. China launched the 5G technical R&D test in January 2016 and the first-stage technical test was finished in September 2016, when the feasibility of the single point key technology was verified. In September 2017, the test result of wireless part (the second stage) was released and the 5G integration solutions of all manufacturers all met the key indicators set by ITU, covering peak rate, time delay, the user's connection capabilities and data traffic density, laying foundation for commercial use of 5G technology in China.

5. **China's optical communication takes the lead in some markets.**

In 1995, China successfully developed the first SDH equipment and ATM switch of 155 Mbps, breaking the monopoly of foreign equipment in the domestic

[3]Source: *Annual Report of Huawei, 2017,* http://www-file.huawei.com/-/media/CORPORATE/PDF/annual-report/annual_report2017_cn.pdf?la=zh.

market. In recent years, Chinese businesses have made substantial breakthroughs in 100G OTN/DWDM and PIN/IP RAN, and 10G EPON/GPON. In 400 GB–1 TB technology, China's businesses first caught up and then took the lead in some areas. Some technical achievements have been made in cutting-edge technologies like super-speed, super-capacity and ULH optical transmission, software-defined elastic optical networks, next-generation optical access to network and silicon photon integration. China's self-supply capacity in key components and materials like optical fiber performs has been improved and its optical communication market scale is increasing, with more and more product catalogues and increasing technical strength and industrialization capability. In 2015, optical fiber manufacturers like FiberHome announced that they had made breakthroughs in low-loss large effective-area fiber technology and released products applicable to 100G/400G backbone networks. They are leaders in the optical cable market for high-end backbone networks and their technology is close to or reach the advanced technology of the world. On the other hand, China is not strong in optical communication chip technology and Chinese optical component manufacturers' technology and R&D capacity still lag behind, so the country has to increase its support and input in the R&D of photonic chips.

6. **Leading breakthroughs have been made in quantum communication.**

Since 2016, guided by the national strategic planning, thanks to the joint-efforts in the field, China has witnessed a series of key achievements in quantum communication. In August 2016, the first quantum satellite "Mozi" was successfully launched for the purpose of scientific experiment and it completed the experiments of entangled quantum distribution, quantum key distribution and quantum teleportation. In May 2017, the light quantum computer was invented jointly by University of Science and Technology of China, Chinese Academy of Sciences—Alibaba Quantum Computing Laboratory, Zhejiang University and Physics Institute of Chinese Academy of Sciences. In September 2017, Beijing—Shanghai Trunk, the world's first quantum confidential communication network, was put into use. China takes the lead in quantum communication research and has launched the construction of quantum confidential communication backbone networks and pilot projects of quantum communication technology application like the cross-border RMB receipt and payment information management system in the fields of national defense, finance and governmental networks.

7. **The Beidou Navigation Satellite System is accessing to global network.**

The Beidou Navigation Satellite System is a global navigation satellite system exclusively developed by China. It is the third of its kind in the world, following the Global Positioning System (GPS) of the United States and GLONASS Navigation Satellite System of Russia. The first Beidou satellite was launched in 2000. Some began to provide service for most places of the Asia-Pacific Region in 2012. The *Mid and Long-term Development Plan for the National Satellite Navigation Industry* released in 2013 proposed that Beidou compatible satellite navigation function should be a standard configuration of cars and smart mobile phones. According to the *National Information Plan for the 13th Five-year Plan Period*,

Beidou system construction and application initiative would be launched in 2017. In August 2016, among the over 3,500 kinds of mobile phone applying for network access licenses, 40% of them support Beidou positioning function.[4] In 2016, the total output value of China's satellite navigation and location-based service industry amounted to RMB 211.8 billion *yuan*, of which 80.8 billion *yuan* was the core output value of the industry directly related to satellite navigation, covering chips, components, algorithm, software, navigation data and terminal equipment. Beidou contributed 70% to the core value of the industry, used in transportation and communication, vehicle and ship monitoring, power industry, farming, fishing, public security and forestry.[5] In November 2017, China successfully launched two Beidou No. 3 satellites, marking that the country's Beidou Navigation Satellite System is accessing to the global network.

8. **The construction of submarine fiber-optic cables connecting the world is being accelerated.**

As a key contemporary international communication means, the submarine fiber-optic cable supports 90% of international communication service; therefore as a major carrier of global information communication, it has become a new sphere of China's optical fiber industry. From 1989 to 1998, the country participated in the construction and investment of 18 international submarine fiber-optic cables. In February 1996, China-Republic of Korea (ROK) undersea fiber-optic cable was completed and opened for use; in November 1997, FLAG was completed and put into operation, the first cross-continental fiber-optic system connecting China, who had participated in the construction. The part in China is 622 km. In September 2000, Shanghai Landing Station of the South-East Asia–Middle East–Western Europe 3 (SEA–ME–WE3) was opened, marking the official opening of SEA–ME–WE3 connecting 33 countries and regions. China Telecom participated in its construction. In June 2017, China Unicom, Cameroon Telecommunications and Huawei Marine Networks Co., Ltd. jointly signed the agreement on SAIL construction. When SAIL is completed, it will be the submarine cable route with the largest transmission capacity and the shortest time delay.

2.2 Multi-link Coordinated Breakthroughs Have Been Made in Integrated Circuit Technology

Integrated circuit (IC) industry, a strategic, basic and leading industry, is a key support for cultivating strategic emerging industries and developing information economy, so it plays a core role in the field of information technology. To facilitate

[4]2016, *White Paper of China's GNSS & LBS Industry*, GNSS & LBS Association of China. http://www.glac.org.cn/index.php?m=content&c=index&a=lists&catid=17.

[5]2016, *White Paper of China's GNSS & LBS Industry, GNSS & LBS Association of China.* http://www.glac.org.cn/index.php?m=content&c=index&a=lists&catid=17.

the development of the IC industry, China launched the *Outline of the Facilitation of the National Integrated Circuit Industry Development* in 2014. In March 2017, the Strategic Alliance of Integrated Circuit Industry was established in Beijing by 62 leading companies, universities, research institutes and social organizations. Driven by both the market and policies, China's IC technology has improved dramatically and has reached advanced level in some areas. For example, breakthroughs have been made in 22 nm technology R&D, 28 nm technology has been applied in mass production, and packaging technology has caught up with the world's level.

1. **China's IC design capacity has reached nanometer level and the design tool is of high reliability.**

Driven by CAD (Computer Aided Design) application development, progress has been made in China's IC design capacity. Reverse design has been transferred to top-down design and super-scale IC design and SoC (System on a Chip) design have been started. Full-custom design has been adopted for some high-demanding e-products. In recent years, Shenzhen Hisilicon, Unigroup Spreadtrum & RDA, SMIC, Spreadtrum Communications and China Huada have emerged as professional IC designers, all of which are of international 15 nm/14 nm design level. China has made some achievements in independent R&D of EDA (Electronic Design Automatic) and Huada Empyrean, in particular, has developed the whole set of EDA tool package, which, though, is not so competitive in the world. Advanced EDA tools are all in the hand of overseas companies. Besides, China's IC IP core (Intellectual Property Core) design technology was developed late. At present, there are only a small number of design types and verified cores. Overseas IP cores are being offered to domestic companies at low prices or for free through cooperation, which hinders the development of China's independent IP core design capability and industrialized application, and hence the construction of SoC design capacity.

2. **China's IC chip manufacturing technology is a key nanometer technology.**

For a long time, China has no property right of its own in terms of IC manufacturing technology, which has resulted in the country's being limited by the West in terms of advanced manufacturing equipment, material and technology, which, in turn, hinders China's development of computers, consumption electronic products, network communications and automotive electronics. Certainly, the situation is turning for the better. In 2013, Integrated Circuit Advanced Process Center of Institute of Microelectronics of Chinese Academy of Sciences made breakthroughs in key leading 22 nm technology research and platform construction, providing technical support for the country's independent R&D of 16 nm and below technology. China is putting the 16 nm/14 nm manufacturing technology into production and speeding up the R&D of 10 nm technology and pre-R&D of 7 nm and 5 nm technologies. SMIC is the only pure-play foundry on the Chinese Mainland that can provide 28 nm processing. In 2016, the manufacturer's operating income from 28 nm technology accounted for 1.6% and was expected to reach 10% in 2017.[6]

3. **Packaging and test technologies are expanding into the mainstream areas of the world.**

Supported by the "Ultra Large Scale Integration Circuit Manufacturing Equipment and Complete Technology", a major special national science and technology project, China's packaging and test capacity and technical level keep improving. A number of high-end packaging technical projects and integrated packaging technological development projects have passed special check and acceptance. High-density packaging technology corresponding to 40 nm/28 nm technical nodes, high-power component packaging technology and packaging system integrated technology have been successfully developed or put into mass production, which have boosted the development of packaging and test industry chains of IC. China's packaging technologies with independent intellectual property right have been recognized by the world. By September 2017, Chinese companies had obtained 585 national/international patents, providing hundreds of packaging and test technical services for hundreds of businesses. During the 13th Five-year-plan Period, China will facilitate the mass production capacity of packaging technologies covering flip chip, solder bump packaging, WCSP (wafer chip scale packaging), and launch new-type packaging technology development covering TSV (through silicon via) and 3D stacked packaging.

2.3 Large-Scale Computing Technology Is Developing Fast to Catch Up with the Highest Level of the World

China is following the global development pace in emerging technical fields like cloud computing, big data and AI, and has launched a large number of R&D activities. Backbone corporations of cloud computing have made breakthroughs in key areas such as large-scale concurrent processing, mass data storage, and energy saving of data centers. Some indicators have reached the world's advanced level. In terms of super-scale data warehouses, distributed storage and computing, and AI-based big data analysis, China is already the leader in the world. Since it has boosted the robot's capacity of perception, control and human-computer interaction, AI has been rapidly upgraded and laid out as an important form in "Internet+" and "Made in China".

[6]Source: *Annual Report of SMIC,* 2016, http://www.smics.com/attachment/2017042716500100032793270_tc.pdf.

1. **China can independently develop self-controlled super-computers, which are advanced in the world.**

The research and development of the super-computer in China, started in 1960s, has experienced three stages. The first stage lasted from the end of 1960s to the end of 1970s, when the research was centered on the parallel processing technology for mainframe computers. The second stage was from the end of 1970s to the end of 1980s, when the research on the vector machine and parallel processing system was done. The third stage lasted from the end of 1980s to today, when MPP and workstation cluster system have been the focus of the research. Thanks to scores of years of efforts, China has achieved great success in the research and development of high-end computer systems. For example, Galaxy, Sugon, Sunway and Shenteng have been developed as high-end computer systems and thus the country has become the third of the world capable of producing high-end computer systems, following the United States and Japan. In June 2017, at the Supercomputing Conference of the world, Sunway TaihuLight, which was developed independently by China, was the first on the list of Top 500 for the second time, thanks to its peak value computing capability of 1,250 trillion per second and its sustaining computing capability of 930 trillion. Galaxy No. 2 super-computer ranked second for the third time at the conference.

2. **The independent R&D capability of cloud computing technologies is being strengthened and breakthroughs have been made in some critical areas.**

China's backbone corporations of cloud computing have made breakthroughs in some critical areas like large-scale concurrent processing, massive information storage and data center energy saving. Some indicators are of the world's advanced level. In early 2017, the world's first cloud server system with 100 million simultaneous user sessions developed by Sugon was put into mass production in Tianjin, so that data centers using traditional infrastructure service can have greater computing capability with the existing hardware to meet such a great number of simultaneous user sessions. China's cloud service manufacturers have developed restart-free hotfix technology. For example, the hotfix technology of UCloud can help to fix all kernel code while re-start is not necessary, and control the outage within 10 ms during the fix. Till now, hotfixes have been adopted for approximately 50,000 times.

3. **Internet businesses are leaders in technologies and all kinds of cloud services are thriving.**

As for basic equipment, Chinese businesses have made breakthroughs in core technology and product R&D in the fields of software and hardware of EB-level storage system and server system with 100 million simultaneous user sessions. In 2011, Baidu, Alibaba and Tencent jointly launched the "Scorpio Project". They started to cooperate with basic telecommunication operators and worked out solutions to high-density enclosed rack servers, occupying major domestic markets and producing positive influence on international markets. Cloud services platforms of Alibaba Cloud and Tencent Cloud, with independent intellectual property right, have entered international markets, and the whole cloud industry of China is developing fast. By October 2017, Alibaba Cloud ecosystem had over 5,000 partners and over 6,000 cloud market products.[7]

4. **Internet businesses have become incubator platforms of big data innovation.**

All Internet businesses are the major incubator platforms of big data innovation. Tencent's distributed large-scale data warehouse supports offline storage and computing of 100 PB-level data, providing massive, efficient and stable big data platform support and policy support. Lenovo has been developing the technology in fields of virtual and distributed storage for over ten years, having successfully applied for more than 90 international technological patents. It has made use of big data-driven active operation, maintenance and system optimization and provided advanced cloud services for hundreds of governmental agencies and corporations. Big data service providers for media make analysis of big data of Internet media through AI natural languages to monitor and assess influence diffusion and hence help to make relevant decisions.

5. **Big data center construction has become the investment hot spot.**

Many regions of China have strengthened the construction of big data industry. The National Information Economy Demonstration Zone has been established in Zhejiang, and national big data comprehensive experimental areas have been set up in Guizhou, Beijing-Tianjin-Hebei Region, the Pearl River Delta, Shanghai, Henan, Chongqing, Shenyang and Inner Mongolia to promote the construction of big data centers. In 2015, the construction of Guiyang-Gui'an Big Data Industry Development Cluster Area was approved by the Ministry of Industry and Information Technology and it is expected that the National Data Storage Disaster Recovery Base and National Cloud Computing Application Base will have been built by the end of 2017. Thus a big data industry cluster will be formed and the total industrial scale is expected to reach RMB 30 billion *yuan*. In Shenzhen, the big data industry is taken as a focus of support, attracting the National Supercomputing Center (Shenzhen Center) and Alibaba Cloud Shenzhen Data Center to be located there to promote the innovative development of the local economy. In Shanghai, the two focuses are the improvement of governmental administration and public service capability and the transfer of the economic development mode. There the big data industry chain of "transaction centers + innovation bases + industrial funds + development alliances + research centers" (five in one) is developing in an environment-friendly way to contribute to establishing the National Data Science Center, the Asia-Pacific Data Exchange Center and the Global Data Economy Center there.

[7] *On the Way to Becoming the World's No. 5 Economy: decoding Alibaba Cloud ecosystem*, http://www.sohu.com/a/201233452_123988.

2.4 AI Brings About New Growth Points

1. **AI technology R&D has seen many achievements.**

In July 2017, the State Council issued *Planning for the New-generation Artificial Intelligence Development*, which proposes the guidelines, strategic goals, key tasks and guarantee measures for the new-generation AI development by 2030. By far, Nanjing University, Tsinghua University and Shanghai Jiao Tong University have achieved a series of academic results in the core AI theory research, and Internet businesses have achieved a series of results in application of human–computer interaction techniques like computer vision and natural language processing. In terms of recognition technology R&D, Baidu has achieved the face recognition accuracy rate of 99.84% and speech recognition accuracy rate of 95% through the in-depth learning network; Tencent has won the championship in the one-million-level face recognition test on the MegaFace, an internationally authoritative face data base; and Alibaba Cloud's ET face recognition technology has its recognition rate of 99.5% in the out-door face detection database. Huawei has set up Noah's Ark Lab to engage in AI machine learning and data mining, with an annual investment of over RMB 50 billion *yuan*. iFlytek, the biggest listed corporation of intelligent speech and AI in the Asia-Pacific Region, has started from speech recognition and then engaged in perception AI and cognition AI. As for AI chip R&D, Cambricon incubated in Chinese Academy of Sciences issued the new-generation of AI chips of the world in November 2017, including Cambricon series of 1H8, 1H1b and 1 M, which can be used in many AI applications. In June 2017, on the list of 50 Smartest Companies announced by *MIT Technology Review*, there are nine Chinese ones.

2. **AI technology is promoting the acceleration of intelligent robot R&D, production and application.**

AI technology is promoting robot technology to be improved in perception, control and human-computer interaction capacity. Robots are used in unmanned working, sophisticated operation, portable collection and cloud-client combination to improve efficiency and practicability. In 2016, 87,000 industrial robots China has significantly expanded its leading position as the biggest market with a share of 30% of the total supply in 2016 (27% in 2015). With sales of about 87,000 industrial robots in 2016—an increase of 27% compared to 2015—China came close to the total sales volume of Europe and the Americas together (97,300 units). Chinese robot suppliers continued to expand their market share from 25% in 2013 to 31% in 2016. Between 2011 and 2016, total supply of industrial robots increased by about 31% per year on average,[8] in comparison with that of the same period of

[8]Source: World Robotics 2017 Industrial Robots, International Federation of Robotics (IFR), https://ifr.org/downloads/press/Executive_Summary_WR_2017_Industrial_Robots.pdf.

the previous year, higher than the increase of industrial robots. With the technological progress and upgrading and transformation of manufacturing and service industries, China will enjoy a bigger market of the intelligent robot industry.

2.5 Emerging Cutting-Edge Technologies Make Internet Application Promising

1. **Virtual reality technology enjoys great potential for development.**

Virtual reality (VR) technology has been listed as a key technology to be developed in the *Planning for Informatization during the 13th Five-year Plan Period* and in the initiatives of "Made in China 2025" and "Internet+", and will be supported by Ministry of Industry and Information Technology, National Development and Reform Commission, Ministry of Science and Technology, Ministry of Culture and Ministry of Commerce. All provincial governments have started to build industrial parks and laboratories and promote the local VR industry. By the end of 2016, nearly 20 provinces or municipalities had planned to develop the VR industry. So far, VR technology has been applied extensively in China. The application is divided into industrial application and public application, the former covering urban planning, education and training, cultural relic preservation, medical care, real estate, exploration and surveying and mapping, manufacturing and aerospace industry, and the latter covering games, socialization, film and TV, live streaming and so on. To be short, the "VR/AR+" era has arrived.

2. **Exploration and R&D of blockchain technology has just started.**

Blockchain technology has attracted nationwide attention. Relevant researches have been launched in that field. In January 2016, the People's Bank of China proposed to define digital currency issuing as the strategic goal, to speed up the research into the application of blockchain technology in financial fields like real-time settlement, resources utilization and sensitive data protection, and to pay close attention to security risks and challenges. In October 2016, the Ministry of Industry and Information Technology released the *White Paper of China's Blockchain Technology and Its Application and Development* (2016), which, for the first time, proposed the standard route for China's blockchain. At present, Chinese corporations are strengthening the R&D of core technologies, such as consensus mechanism, programmable agreement, distributed storage and digital signature. By June 2017, China's State Intellectual Property Office Database had published 34 effective patents containing the key word "blockchain".[9]

[9]Source: China's State Intellectual Property Office Patent Database, http://www.pss-system.gov.cn/.

Chapter 3
Digital Economy as the New Engine for China's Economic Growth

Chinese Academy of Cyberspace Studies

General Secretary Xi Jinping confirmed in the Report of the 19th CPC National Congress that "China's digital economy is thriving". Digital economy is the representative of the country's cultivating new driving forces through informatization to promote new development. According to Digital Economy Development and Cooperation Initiative released at Hangzhou Summit of G20, digital economy refers to a series of economic activities, in which digitalized knowledge and information are taken as key production elements, the modern information network as an important carrier and the effective use of information communication technology as the drive for efficiency improvement and economic structure optimization. China is promoting digital industrialization and industrial digitalization, having achieved a lot in the fields of e-commerce, information service, network terminating equipment, Internet finance, sharing economy and "Internet+" integration. The country's digital economy scale in 2016 reached RMB 22.58 trillion *yuan*, ranking second in the world, with a growth rate of 21.2%. Digital economy accounts for 30.3% of the national GDP,[1] and has become the new engine for and new highlight of the country's economic growth.

[1]Source: Chinese Academy of Information and Communications Technology, *White Paper of China's Digital Economy Development (2017).*

Chinese Academy of Cyberspace Studies
Beijing, China

Chinese Academy of Cyberspace Studies (ed.), *China Internet Development Report 2017*,
https://doi.org/10.1007/978-3-662-57521-5_3

3.1 Network Terminal Equipment Manufacturing Is Being Upgraded Fast

3.1.1 China's Smartphones Are Competitive Products on the Global Market

1. **China's smartphone market has become stable.**

Smartphone business was launched in China in 2010. It has witnessed fast development in the past seven years. China-made mobile phones are the leaders on the domestic smartphone market. The top four smartphone sellers in the first half of the year 2017 were all Chinese businesses, with Huawei's sales accounting for 21.6%, ranking first, followed by that of OPPO, vivo and Mi,[2] while Samsung and iPhone witnessed the decrease of their sales. China's smartphone market has become stable, with the number of currently owned phones amounting to 106 million in 2016, which is decreasing steadily. The future market will be gradually become saturated and the increase will be stable. The number of currently owned smartphones in China from 2011 to 2017 is shown in Fig. 3.1.

2. **China-made smartphones are occupying the global market.**

In recent years, through fierce competition, China-made smartphones are more and more competitive and influential on the global market, with their international market shares increasing steadily. They are a miniature of China's high-end manufacturing's access to the international market. In 2016, the shipments of Chinese smartphones totaled 465 million, accounting for on third of the global total 1.5 billion.[3] Leading brands of smartphones made in China are competitive with famous international ones. Thus Chinese smartphones' overseas sales have witnessed great success. ZTE smartphones are sold in 160 countries and regions and the corporation has established cooperation with 230 operators in the world. Huawei, based on its Hisilicon chips, has pushes its smartphones to develop towards high end. In 2016 its smartphone shipments increased by 21%, much higher than that of Samsung and iPhone; OPPO and vivo aim their products to fashion groups in cities and their smartphone shipments in 2016 witnessed an increase of 109% and 78% respectively. The global market share of China-made smartphones from 2012 to 2017 is shown in Fig. 3.2.

[2]Source: iiMedia Research, *Research Report on the Smartphone Market in the First Half of 2017*.

[3]Source: Counterpoint Research, *Research Report on Global Mobile Phone Market (2016)*.

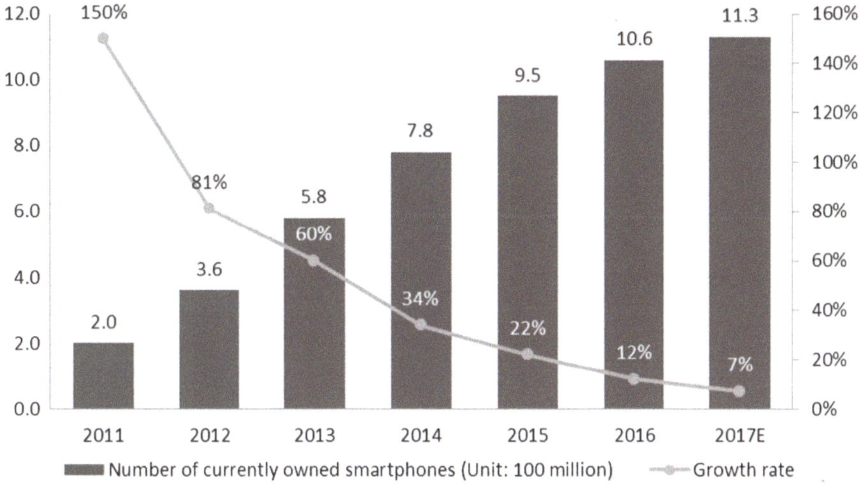

Fig. 3.1 Number of currently owned smartphones in china from 2011 to 2017 (*Source* chinaidr.com)

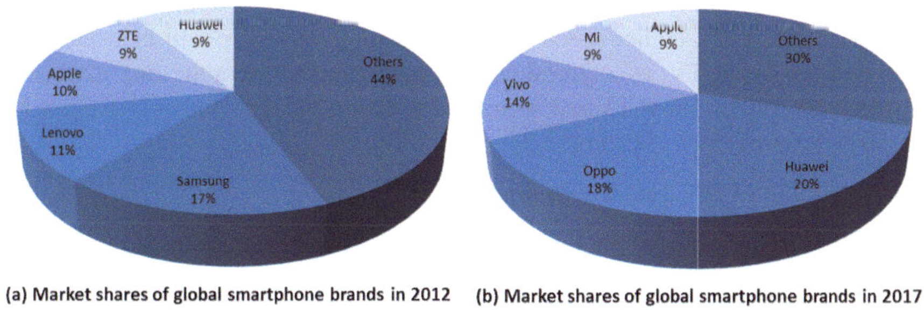

Fig. 3.2 Global market share of chinese-brand smartphones (2012 and 2017) (*Source* Sina Finance)

3.1.2 *New-Type Intelligent Terminals Are Thriving*

The rapid development of new technologies is leading to the fast growth of intelligent terminals and fast product replacement in the fields of smart home, IoV, AI and VR.

1. **The ecosystem of smart home is being established.**

China's smart home industry is witnessing rapid development in recent years, attracting businesses of traditional hardware, Internet, real estate and house decoration. Baidu, Huawei, Mi and Lenovo are increasing their investment to integrate AI, speech recognition and deep learning into smart home. With the commercial application of mobile communication technologies such as 5G, Bluetooth 5.0 and

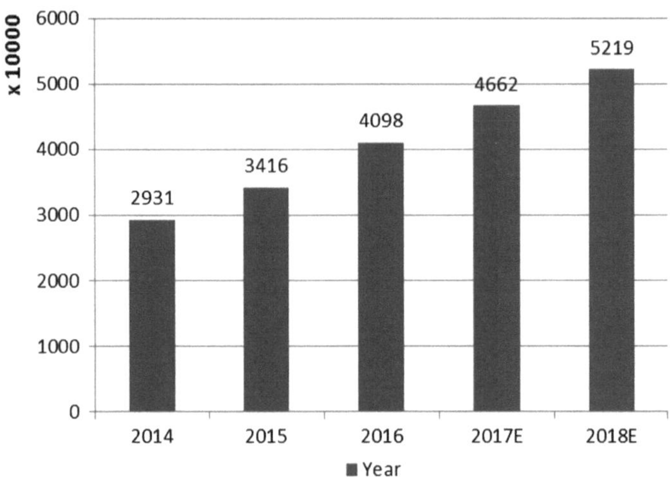

Fig. 3.3 China's smart TV set sales and expected sales (2014–2018) (*Source* iiMedia Research, Special Research Report on China's Smart TV Set Market)

next-generation Wi-Fi, smart home industry will witness new reforms, its products will be released at a higher speed and its industrial ecosystem will be established step by step. Smart TV, as a typical area of that field, will see a thriving market. In 2016, 40.98 million smart TV sets were sold in China and the number is expected to be 50 million in 2018 (See Fig. 3.3).

2. **Terminals of IoV are developing fast.**

Internet businesses and chip manufacturers all take the car operating interactive system and the car processor as the new breakthrough points and have entered the market of IoV. Traditional auto manufacturers have launched their strategy for IoV. For example, BMW succeeded in 100% networking of all its models sold in China since June 2015; General Motors (Shanghai) plans to realize that goal by 2020. The LTE-V2X standard of IoV led by China has become mature and the standardization was approved in March 2017 by 3GPP, the international telecommunications standards association. By August 2017, China Mobile, China Unicom and China Telecom had had over 58.06 million users of their IoV and 683 car networking terminals had obtained the access licenses from the Ministry of Industry and Information Technology. Among them, there are 101 4G terminals.[4]

3. **China's AI industry is witnessing an explosive growth.**

China's AI industry was started later than that of some countries, but with the involvement of giant companies of science and technology and manufacturing and

[4]Source: Chinese Academy of Information and Communications Technology, *White Paper of Internet of Vehicles* (2017).

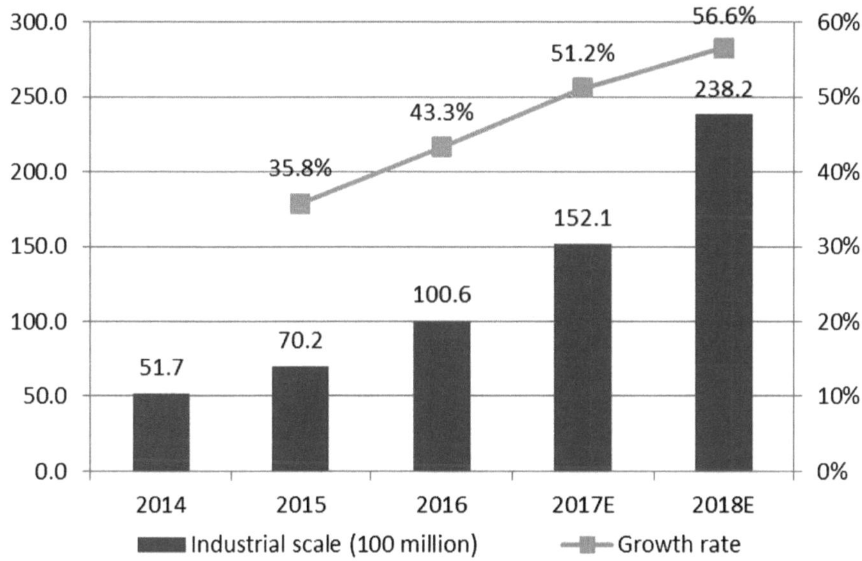

Fig. 3.4 China's AI industrial scale and expected scale (2014–2018) (Source iiMedia Research, Special Research Report on China's AI Industry (2017))

with the founding and growth of a number of startups in vertical areas, an increasing number of industrial and consumer application products are emerging. In 2016, AI application innovations saw an explosive growth in China. In March that year, Chinese Academy of Sciences released the world's first neural network processor Cambricon; in July, 4 Paradigm, an AI financial service provider, launched its development platform named "Prophet"; in August, Alibaba released its AI named ET; in September, Baidu showed its Baidu Brain; in October, Ant Financial and Tsinghua University jointly founded an AI laboratory; in December, Tencent Cloud released to the world seven AI services, trying to keep up with the leading level of the world in sub-segmentation. In 2016, China's AI industrial market scale was over RMB10 billion *yuan*, an increase of 43.3% in comparison with that of the same period in the previous year. It is expected to amount to RMB23.8 billion *yuan* in 2018 (See Fig. 3.4).

4. **China's VR application is progressing fast.**

Despite its small scale in general, China's VR applications market is growing fast, represented by VR-TIME, Baofeng Mojing, WhaleyVR, uSens and Noitom. China's VR market scale in 2016 was RMB3.46 billion *yuan* and its annual compound growth rate in the coming five years is expected to be over 80%. The total scale will be more than RMB 10 billion *yuan* in 2018 and 79 billion *yuan* in 2021. China will probably be the largest VR market of the world. The largest-scale segmentation will be the VR headset, whose scale of RMB2.05 billion *yuan* accounted for 59.2% of the total in 2016. Then came VR Experience, whose market

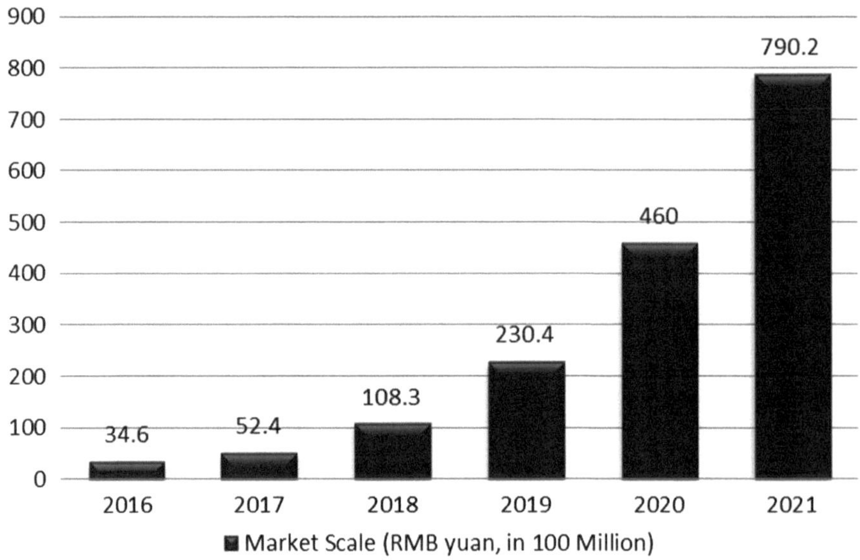

Fig. 3.5 China's VR market scale and expected scale (2016–2021) (*Source* Research Report on China's VR Industry, iiMedia Research and GreenlightInsights)

scale is expected to be RMB5.2 billion *yuan* in 2021. The annual compound growth rate of other VR hardware like the motion capture system, touch control and sound collection equipment is expected to be 54.6% and the total scale will be RMB 3.6 billion *yuan* in 2021[5] (See Fig. 3.5).

3.1.3 The Trusted System of Network Equipment Is Being Established

1. **China's independent chip industry enjoys good momentum.**

During the 12th Five-year Plan Period, the supporting policies for IC industry were consolidated. In particular, the establishment of the national Integrated Circuit Industry Development Investment Foundation has contributed to the fast development of China's chip industry and thus the country has become one of those with the fastest development in that area. China's sales revenue of the chip industry witnessed an annual growth of 20% from 2013 to 2016 and the scale of that industry in 2016 amounted to RMB 433.6 billion *yuan*.[6] The capacity of chip

[5]Source: *Research Report on China's VR Industry: Market Data*, jointly released by iiMedia Research and GreenlightInsights.

[6]Source: China Semiconductor Industry Association, www.csia.net.cn.

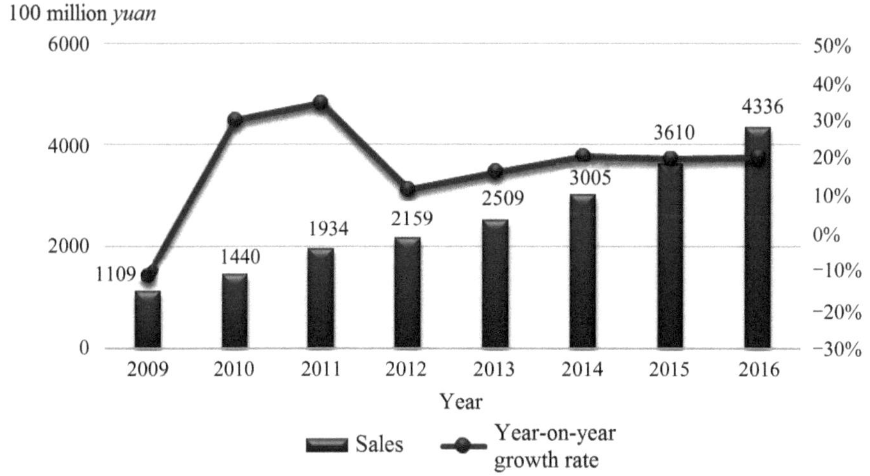

Fig. 3.6 China's integrated circuit industry sales from 2009–2016 (*Source* China Semiconductor Industry Association)

manufacturing on Chinese Mainland accounts for 10.8% of the world's total. Nine 12-inch wafer production lines have been built and another eight are being built[7] there. China's leading corporations in that area are reinforcing their chip R&D. Mobile terminal processor chips of Hisilicon and Spreadtrum can meet the demand for use. Tongfang Guoxin, Unigroup Guoxin, Wuhan Xinxin and Fujian Jinhua have invested in founding chip manufacturers and acquired upstream and downstream corporations in the chip industry chain. Corporations like China Electronics are constructing the industrial ecosystem centering around chips, having formed the complete layout made up of chips and complete hardware machines and then terminal equipment, sensors, communication networks, operation systems, platforms and applications. In the future, China's chips will be more competitive on the global market. The country's IC industry sales from 2009 to 2016 are shown in Fig. 3.6.

2. **China's sensor industry enjoys steady development.**

China started to develop the sensor industry during the 7th Five-year Plan Period. After 30 years of steady development, a complete industry chain has been formed. At present, the country is trying to make breakthroughs in the core technology and manufacturing techniques concerning sensors, accelerating the application of sensors in industrial production, medical diagnosis, ocean survey and environmental protection, and pushing the sensor industry to develop towards high end and to be intelligent and miniaturized. In 2016, China's sensor market scale

[7]Source: IC Insight, *Prediction Report on the Global Wafer Capacity 2007–2021.*

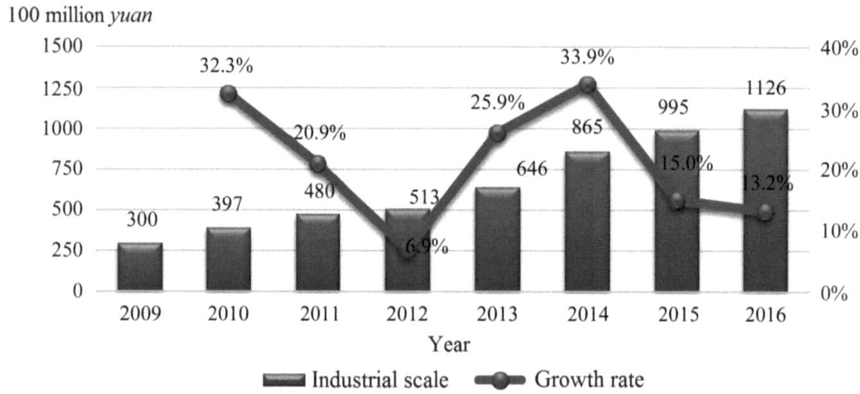

Fig. 3.7 China's sensor manufacturing market scale from 2009–2016 (*Source* ibaogao.com, Consulting Report on Operation and Investment Strategies of China's Sensor Manufacturing 2017–2023)

reached RMB112.6 billion *yuan* (See Fig. 3.7), with a year-on-year growth rate of 13.2% and a steady growth in the future.

3. **China maintains the market advantage in computer equipment.**

In recent years, due to the sluggish market and the rising of the mobile Internet, the output of microcomputers has been decreasing. Lenovo has maintained its leading position as the world's largest PC manufacturer for four years in a row since 2013.[8] There has been a continuous growth in the fields like servers and storage systems. Among the world's top five businesses in server output in 2016, there were three Chinese ones, namely, Huawei, Lenovo and Inspur. Besides, supercomputers represented by Sunway TaihuLight and high-end fault-tolerant computers represented by Inspur TS K1 have been developed, which has boosted the high-end computer industry in China. The country's microcomputer output and growth rate from 2011 to 2016 is shown in Fig. 3.8. The output of the world's top five server businesses in 2016 is shown in Fig. 3.9.

4. **China's security protection equipment industry is rising.**

After thirty years of development, China's security protection equipment industry has transferred from analogy and digitalization to intelligentization. Within the country, as the demand for safety cities and civil safety protection is increasing, the industry of intelligent safety protection equipment dominated by video surveillance equipment has become an important part in infrastructure construction.

[8]Source: Gartner's *Report on Global PC Sales in the Past Years.*

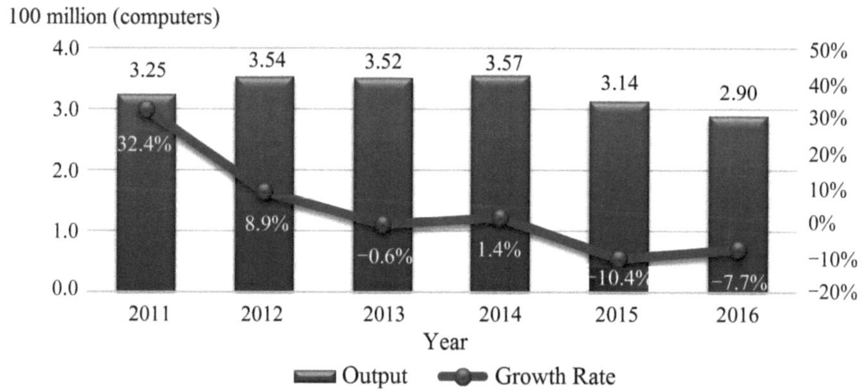

Fig. 3.8 China's microcomputer output and growth rate from 2011 to 2016 (*Source* CCID, February 2017)

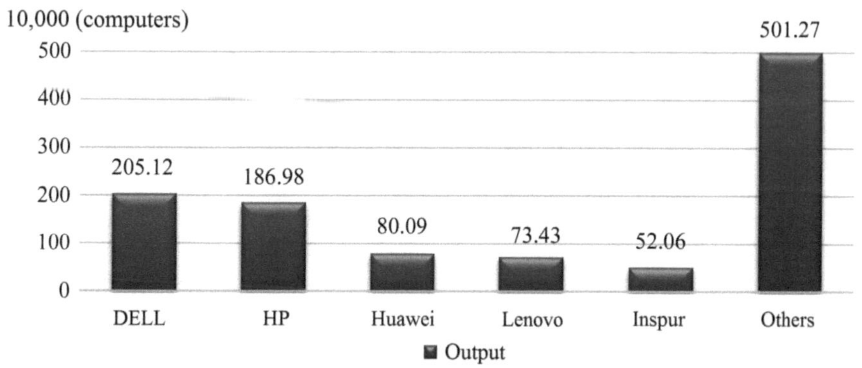

Fig. 3.9 Output of the world's top five server businesses in 2016 (Gartner's Market Monitor Data of 2016)

In 2016, China's safety protection market scale reached RMB540 billion *yuan* (See Fig. 3.10), with a compound annual growth rate of 15% since 2012. The country's safety protection equipment distinguished itself at G20 Summit and Rio Olympics, showing to the world the leading level of China's safety protection products and technology. Among the world's top 50 businesses of safety protection equipment in 2016, Hikvision ranked first, and Dahua Technology ranked fourth. In the field of video surveillance, Hikvision has ranked first in the world for six years in a row in terms of market occupancy.[9]

[9]Source: Data released by IHS, an international market survey organization.

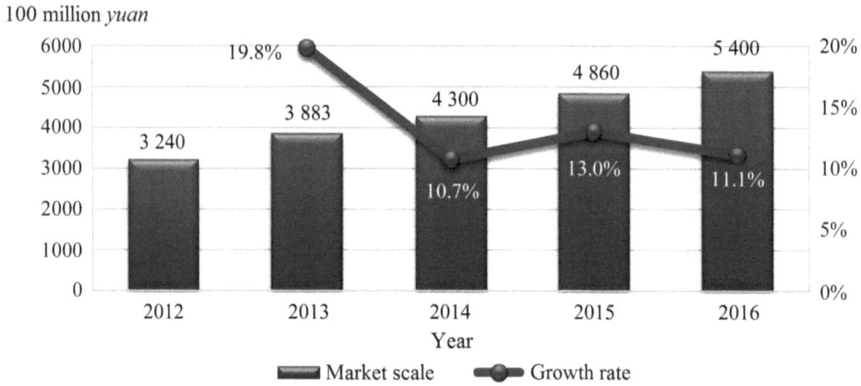

Fig. 3.10 China's safety protection market scale from 2012 to 2016 (*Source* ibaogao.com, Research Report on In-depth Assessment and Investment Prospect concerning China's Safety Protection Industry from 2017 to 2022)

3.2 Internet Information Service Is Being Rapidly Industrialized

With the implementation of the "Broadband China" strategy and the construction of 4G networks, China's information consumption is rising, its scale increasing from RMB2.2 trillion *yuan* in 2013 to 3.9 trillion in 2016, with an annual increase rate of 21%, 2.4 times of the traditional consumption growth rate of the same period. The whole industry contributed 0.26% to the GDP growth of the country that year.[10] With the formulation of *Guiding Opinions on Expanding and Upgrading the Information Consumption and Keeping Releasing the Domestic Demand Potential* by the State Council, Internet information service will witness a new round of high-speed growth.

3.2.1 The Cyber Content Industry Is Prospering

1. **The online entertainment industry is rising rapidly.**

The online entertainment industry covering online video, music, literature and games is developing comprehensively and rapidly. By June 2017, the number of online video audiences in China had reached 564 million (See Fig. 3.11); that of online music listeners, 524 million (See Fig. 3.12); that of online literature readers,

[10]Source: Tang Guijiang (a journalist from chinanews.com): *Vice Minister of Industry and Information Technology: China's Information Consumption Scale Reached RMB3.9 trillion yuan in 2016*, China News Service from Guangzhou, March 23, 2017.

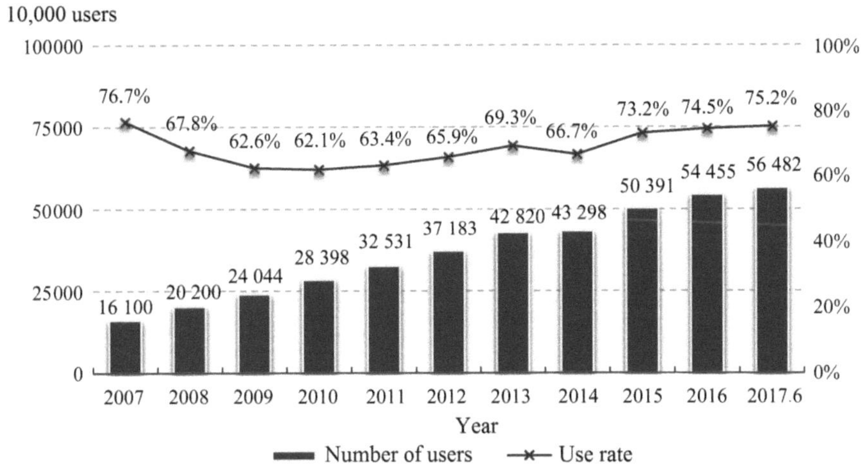

Fig. 3.11 Number of China's online video audiences and use rate (2007-June 2017) (*Source* CNNIC, 39th and 40th Statistical Report on China Internet Development)

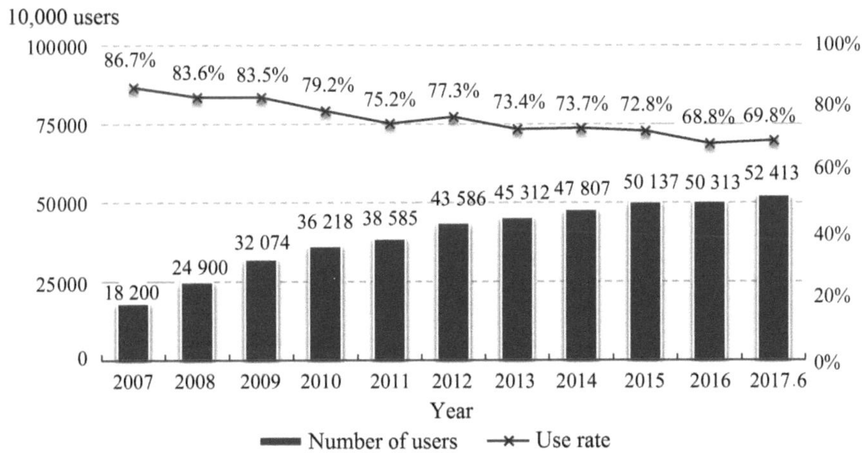

Fig. 3.12 Number of China's online music listeners and use rate (2007-June 2017) (*Source* CNNIC, 39th and 40th Statistical Report on China Internet Development)

352 million (See Fig. 3.13); and that of online game players, 421 million (See Fig. 3.14). In 2016, the market scale of China's online entertainment industry was RMB263.146 billion *yuan*, of which, that of online games was 178.92 billion *yuan*, a year-on-year growth of 24.6%[11] (See Fig. 3.15); that of online video, 60.9 billion

[11]Source: iResearch, *Research Report on China's Online Game Industry in 2016.*

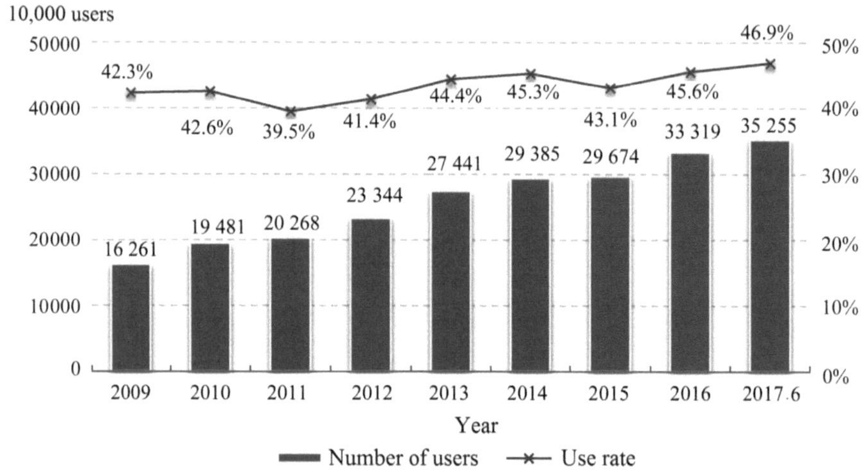

Fig. 3.13 Number of China's online literature readers and use rate (2009-June 2017) (*Source* CNNIC, 39th and 40th Statistical Report on China Internet Development)

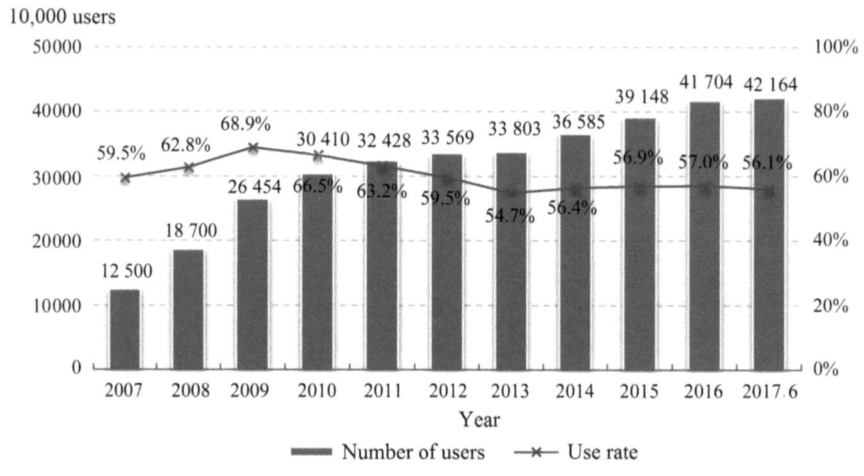

Fig. 3.14 Number of china's online game players and use rate (2007-June 2017) (*Source* CNNIC, 39th and 40th Statistical Report on China Internet Development)

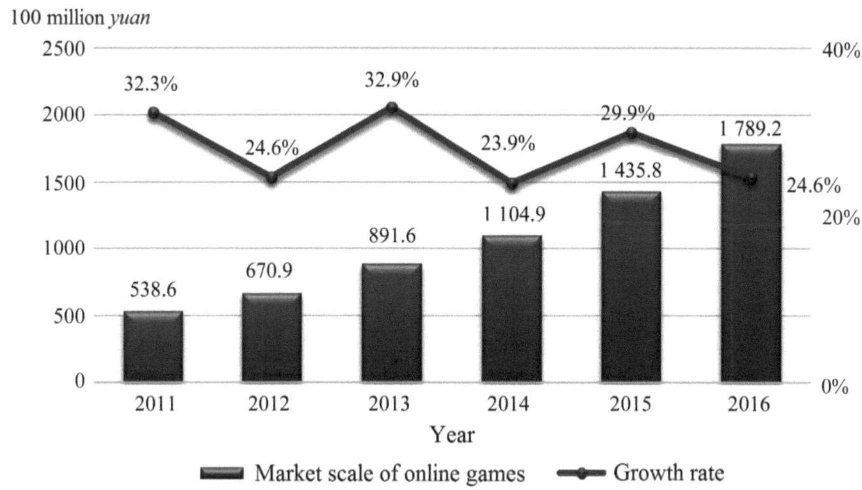

Fig. 3.15 China's online game market scale and growth rate (2011–2016) (*Source* iResearch, Research Report on China's Online Game Industry 2016)

yuan, a year-on-year growth of 55.9%[12]; that of online music, 14.326 billion *yuan*, a year-on-year growth of 39.4%[13]; and that of online literature, 9 billion *yuan*, a year-on-year growth of 28.6%.[14] Different industries of online entertainment are speeding up their integration with each other. A number of businesses represented by Tencent, NetEase and Giant Interactive Group Inc., taking intellectual property originality as the core, are developing the online entrainment industry covering video, live streams, games, literature, socializing and movie ticketing, so that they can enjoy concentrated, integrated and interactive development.

2. **Online advertising accounts for an increasing percent.**

Since 2012, the very first year of programmed purchase, promoted by the sharp increase of the number of Internet users, the time length of digital media use and the online video and audio service, online advertising has witnessed the dramatic growth of its market share while traditional advertising on TV and broadcast and in newspapers and magazines has witnessed the continuous decrease. In 2016, China's online advertising market scale reached RMB290.27 billion *yuan*, with a year-on-year growth rate of 32.9%, accounting for 68% of the total income of five

[12]Source: Development Research Center and Online Video & Audio Program Control Department of State Administration of Press, Publication, Radio, Film and Television of the People's Republic of China, *Report on China's New Video & Audio Media Development (2017).*

[13]Source: School of Music and Recording Arts of Communication University of China, *Report on China's Digital Music Industry Development.*

[14]Source: irinbank.com, *Analysis of China's Online Literature Industry Scale and Status Quo in 2016.*

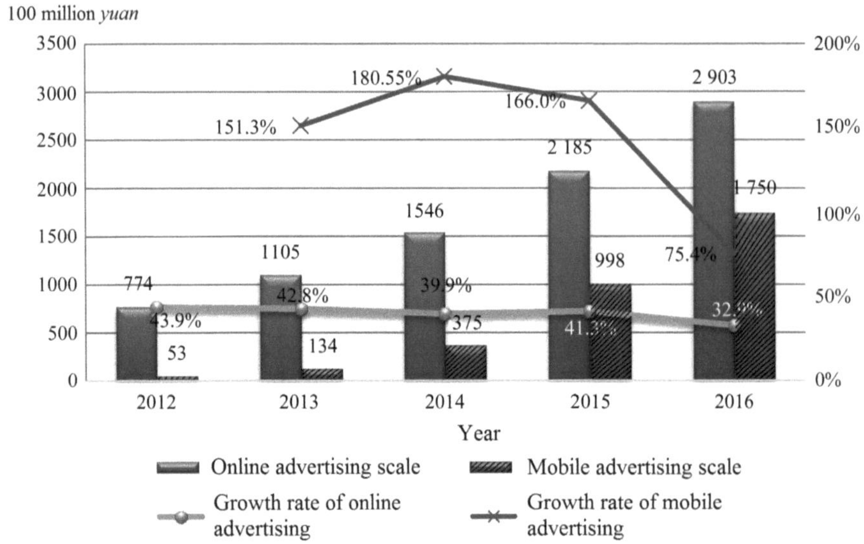

Fig. 3.16 China's online advertising and mobile advertising market scale and growth rate (2012–2016) (*Source* iResearch, Annual Monitoring Report on China's Online Advertising Market (2017))

advertising media forms. The market scale of mobile advertising was RMB175.02 billion *yuan*, with a year-on-year growth rate of 75.4% (See Fig. 3.16), accounting for 60.3% of the total advertising income and exceeding 50% for the first time. Now China's online advertising forms keep being innovated, big data application and analysis capability keep being improved, and feed advertising on socializing, news and video websites keep growing. Simultaneously, the rise of live streaming platforms and the concept of "Internet celebrity" produce influence on the advertising forms and ideas. Interior advertising will become the trend in the future.

Column 4 Programmatic Buying of Online Advertising

Programmed purchase refers to advertising based on the automatic system (technology) and data, aimed to improve the efficiency and effect of digital advertising by utilizing the clients' data and insight to cover certain users at certain time in a certain environment.

Since 2012, programmed purchase has attracted investment and thus hundreds of programmed purchase platforms have emerged. After five years of development, a market system has been formed involving media, advertisement transaction markets, programmed purchasers, monitors, advertising agencies, and advertisers. Traffic scale and quality, technical infrastructure, transaction mode and all parties' awareness have all become mature in that field. However, while giving a full play to technology and data in advertising, programmed purchase has also given rise to problems, such as cheating in the traffic, falsified transaction, and nontransparent advertising process. All these problems occurred together in 2016, so programmed purchase was suspected by advertisers and then the number of transactions witnessed a sharp decrease. Programmed purchase of online advertising is witnessing a transfer period.[15]

3.2.2 *The Search Engine Market Keeps Expanding*

In recent years, China's search engine market has witnessed a steady growth. In 2016, the income of that market was RMB80.5 billion *yuan*, with a growth rate of 4.5%, accounting for 12.5% of the world's total. It is expected to increase in the future.[16] By June 2017, the number of search engine users in China was 609 million, with the use rate of 81.1% (See Fig. 3.17).

1. **There are continuous innovations in search technologies.**

With the emergence of new information technologies, major search engines like Baidu and Sogou have increased their input in cutting-edge technologies like big data and AI and improved their users' experience. Automatic speech recognition, image recognition and video identification enjoy a high success rate, enabling the users to search what they want to and find what they search for. As for automatic speech recognition, the relative error rate of Baidu APP's speech search has decreased by over 25% and its Chinese speech recognition rate is over 90%. The accuracy rate of Baidu's image search has increased sharply from 20% to 80%, and its testing effect is better than that of the Facebook search technology. The daily search volume on Baidu's video search platform is over 50 million. At the same time, AI robot-aided search has become a part of the standard configuration of all major search engines. For example, Baidu's Duer, Sogou's Speed Assistant and Microsoft Bing's Tay are widely recognized.

2. **The search engine service mode is being transformed rapidly.**

The search engine industry is becoming more vertical and specialized. Its commercial mode is becoming diversified and its profit model is transforming from single information search service to comprehensive online service platforms. Big data of APPs of all major search platforms can provide search service covering maps, shopping, life, news and medical care, meeting the diverse demand of the users for news hotspots, commodity consumption and so on. For example, Sogou

---

[15] iResearch, *Annual Monitoring Report on China's Online Advertising Market (2017).*

[16] Source: Analysys Industry Database, *Quarterly Monitoring Report on China's Search Engine Market (Quarters 1–4, 2016).*

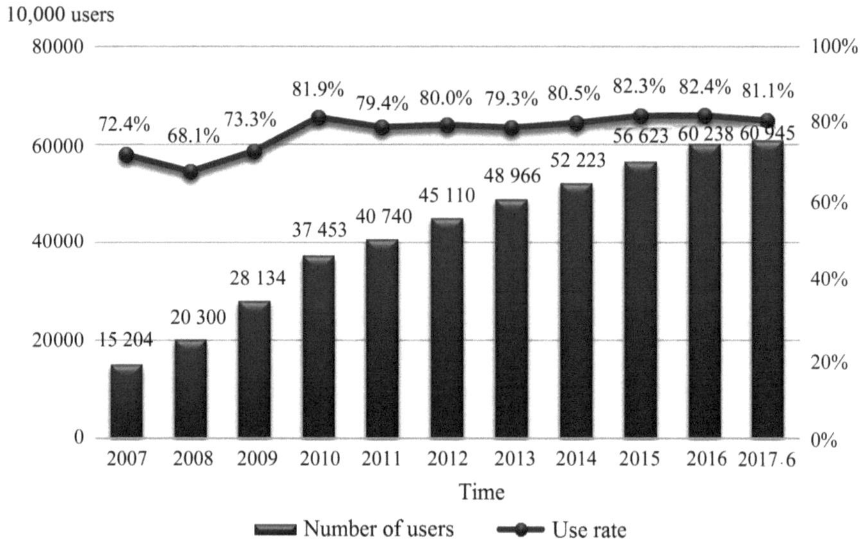

Fig. 3.17 Number of search engine users and use rate in China (2007-June 2017) (*Source* CNNIC, 39th and 40th Statistical Report on China Internet Development)

Search is linked to Wechat, QQ Tribe and Zhihu, and it has reached an agreement on cooperation with Microsoft to optimize the service in socializing, news broadcast, special Q&A, and English and academic search; Baidu has begun to follow the consumption payment senario trend and thus launched direct arrival of goods, local life service, financial service and culture service; so.com and sm.cn have also launched online shopping and O2O life service platforms.

3. **Mobile search has become the major driving force for the market growth.**

Since 2011, the number of mobile search engine users has grown faster than the number of all APP users. By June 2017, the former had reached 593 million (See Fig. 3.18), 1.76 million more than that at the end of 2016, with an increase rate of 3.1%, 1.9% higher than that of the total number of APP users. The use rate of mobile search engines in 2017 was 81.9%, much higher than that of PC terminals. The sales revenue from mobile APPs contributes more and more to the total revenue. In 2016, the total mobile sales revenue of Baidu Search accounted for 63% of the total revenue; the mobile terminal traffic of Sogou Search increased by 70%, contributing 75% to the total traffic.[17]

[17]Source: The *Financial Report of Baidu (2016)* and The *Financial Report of Sogou (2016)*.

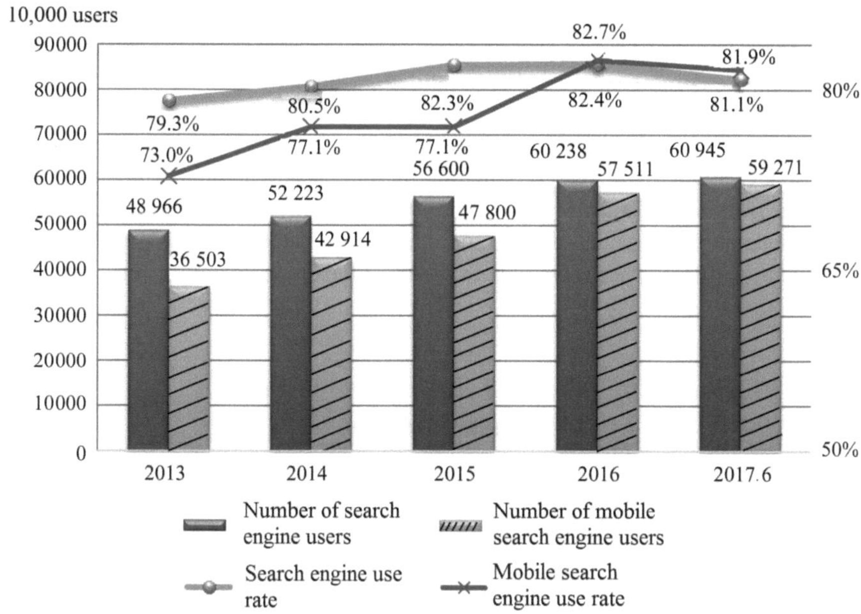

Fig. 3.18 Number of mobile search engine users and search engine use rate (2013-June 2017) (*Source* CNNIC, 39th and 40th Statistical Report on China Internet Development)

3.3 E-Commerce Is Gaining Momentum

E-commerce, a new transaction form, has led manufacturers, circulating companies, consumers and governmental organizations into a new commercial era. The Chinese Government provides strong support for the development of e-commerce, having released documents of policies like *Opinions of the General Office of the State Council on Accelerating the Development of Electronic Commerce*, *Plan for the Development of E-commerce during the 11*th *Five-year Period* and *Opinions of the State Council on Vigorous Development of E-Commerce to Accelerate Development of New Economic Driving Force*, in all of which the strategies, goals and measures are defined clearly. It keeps improving the environment for policies, supervision, market, finance and logistics and thus enables e-commerce in China to have enjoyed two decades of fast development and to have become the most active and representative sector in the digital economy. In 1997, chem.cn, the very first B2B e-commerce platform of China, was launched and one year later, B2B businesses like Alibaba and made-in-china.com were founded. Eachnet.com, the first C2C platform of China, and 8848, the first B2C platform, were put into operation in 1999. Taobao and JD.com, the first group of B2C platforms, rose to the top in 2003,[18] leading to the sustaining rapid growth of online retail turnover and the thriving of mobile e-commerce, e-commerce in rural areas and cross-border e-commerce. In 2016, the number of online shoppers increased to 469 million and the e-commerce turnover increased to RMB26.1 trillion *yuan* (See Fig. 3.19), equal to 35% of the total GDP of the country. Since 2014, the e-commerce turnover has maintained the compound annual growth rate of 31.24%, with 37 million people directly or indirectly employed in that trade, and the annual taxes from it have amounted to RMB 200 billion *yuan*. China is now one of the countries with the most developed e-commerce.

3.3.1 B2C and the Mobile Terminal Have Become the Major Online Retail Platforms

1. **China has the highest online retail volume in the world and diversifying online commodities and retail means.**

Online retail is an indispensable part and development highlight of China's e-commerce. Since 2008, the country's online retail turnover[19] has maintained the compound annual growth rate of 32%. In 2013, it amounted to RMB 1.9 trillion *yuan*, higher than that of the United States for the first time, and thus it became the largest online retailing market. In 2016, China's online retail turnover amounted RMB5.16 trillion *yuan* (See Fig. 3.20), with a year-on-year growth of 26.2%, contributing 30% more to the added value of the total turnover of social consumer goods, with that of physical commodities being RMB4.19 trillion *yuan*, seeing its growth rate 15.2% higher than that of social consumer goods in the same period. Meanwhile, the foundation of the online retailers, commodities and retail means keeps being consolidated, with the commodities and retail means being diversified. The new "retail" has led to the integrated development of online and offline retail, meeting the users' increasing demand for commodities.

Column 5 Double 11 Shopping Carnival

Double 11 Shopping Carnival refers to the online promotion on November 11th. It originated from the promotion launched by Taobao (Tmall) on November 11th, 2009. Though only a small number of retailers did business

[18]Wang Bingnan: Status Quo and Development Trend of E-commerce both at Home and Abroad: 29th *Special Lecture for the Standing Committee of the 12*th *National People's Congress (NPC)*, June 29th, 2017.

[19]Online retail turnover refers to the commodity and service retail turnover obtained through self-built websites and public network transaction platforms.

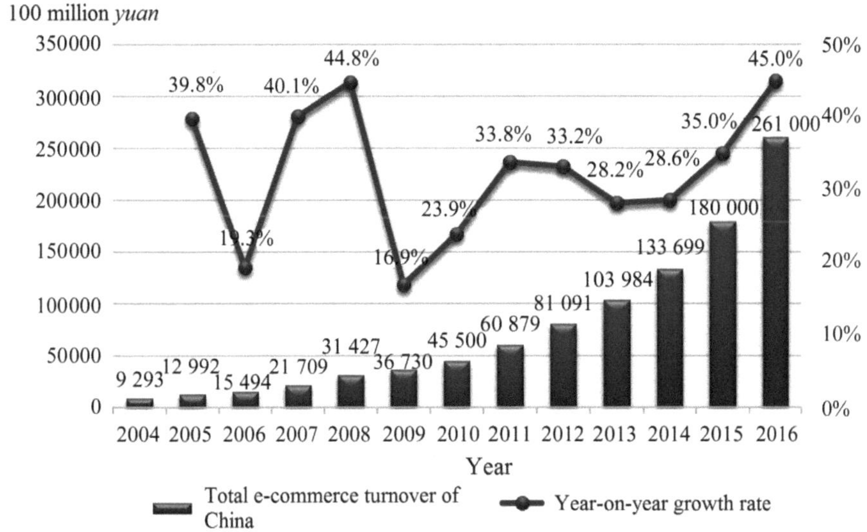

Fig. 3.19 China's total E-commerce turnover and growth rate (2004–2016) (*Source* Report on China's E-commerce made by Ministry of Commerce in recent years)

on the website and promotion strength was limited at that time, the turnover was much higher than expected, which resulted in November 11th becoming a fixed day of Tmall's large-scale promotion. Then other e-commerce platforms, factories and operators followed suit, which led the promotion to become a great yearly event of e-commerce nationwide.

On November 11th, 2014, Alibaba's turnover amounted to RMB57.1 billion *yuan*. On November 11th, 2015, Tmall's turnover was RMB 91.217 billion *yuan* and on the same day next year, the figure was RMB120.7 billion *yuan*, with that of wireless turnover accounting for 81.87%. In the six days between the 11th and 16th of November 2016, the post-offices and express companies collected 1.12 billion express goods and delivered 790 million. On November 11th, 2017, Alibaba's retail turnover was RMB168.2 billion *yuan* and JD.com's accumulative order value was over RMB127.1 billion *yuan*. Double 11 has so far become a worldwide e-commerce carnival, covering 235 countries and regions, from the Greenland in the north to Chile in the south. The explosive growth of the order amount and turnover on November 11th for years shows that online transaction has transformed from a compensation for retailing to a major means boosting the domestic demand, forcing the traditional retailing to upgrade itself.

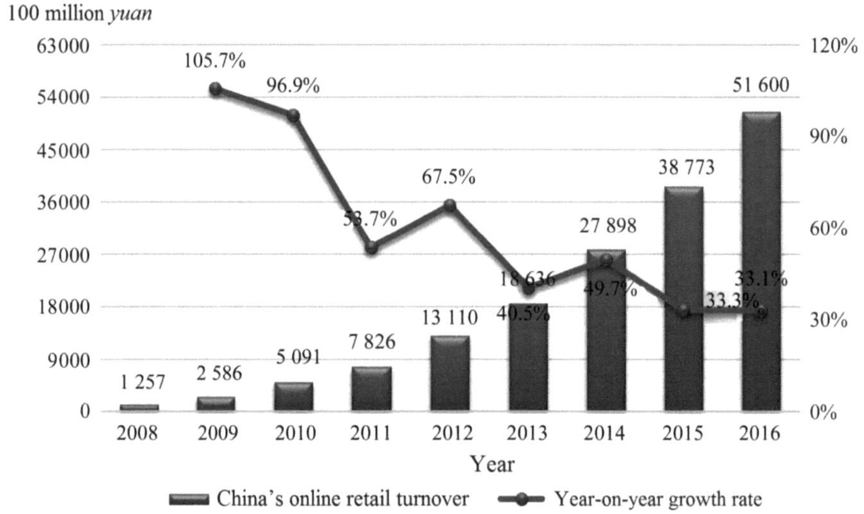

Fig. 3.20 China's online retail turnover and growth rate (2008–2016) (*Source* Report on China's E-commerce made by Ministry of Commerce in recent years)

2. **B2C has become the mainstream of online retailing, with its market share exceeding that of C2C.**

In the early period of China's e-commerce development, C2C enjoyed rapid development and a large scale. But with the improvement of consumption capability, B2C has gradually replaced C2C and become the mainstream. In 2015, the transaction volume on the B2C market amounted to RMB two trillion *yuan*, accounting for 51.9% of the total e-commerce transaction volume, exceeding that on the C2C market for the first time.[20] In 2016, the transaction volume on the B2C market amounted to RMB2.6 trillion *yuan*, with an increase of 31.6% in comparison with that in 2015, much higher than that on the C2C market, which was 15.6%.[21] With the change of the users' consumption demand, transaction volume on C2C platforms has witnessed a decrease year by year while creative B2C modes have emerged one after another. C2B customization, home delivery, crowdfunding, socialized group booking, quality selection and direct supply from origin and periodic supply have emerged, promoting the sustaining growth of B2C retail turnover.

Column 6 Alibaba's Platform E-commerce

Founded in Hangzhou in 1999, Alibaba Group has developed into a large group company with a complete Internet business system, including Taobao,

[20]Source: iResearch, *Report on China's E-commerce Data (2015).*

[21]Source: iResearch, *Report on China's E-commerce Vitality (2016).*

Tmall, Juhuasuan, Aliexpress, Alibaba.com, 1688, Alimama, Alibaba Cloud, Ant Financial, and Cainiao. The Group was listed with New York Stock Exchange on September 19th, 2014 and it announced on April 6th, 2016 that it had become the largest retailing platform of the world.

Taobao and Tmall are typical platform-type e-businesses, providing showcase opportunities and information sources as well as credit guarantee and all kinds of service interfaces while the design, R&D, manufacturing, marketing, storage, delivery and after-sale service are done by the sellers, who form a big market on the platform with varieties of products oriented to the demand. Alibaba e-commerce platforms focus on information flow while its capital flow and logistics are jointly done by Ant Financial and Cainiao. The platforms gather a large amount of information about the market and provide excellent service through value-added service. Adhering to the mission of "making all business easy", Alibaba keeps the platforms open, making the whole system available to entrepreneurs of Internet service while gathering a number of sellers and buyers, and thus creating new business modes and forming a big closed loop of industrial chain.

Column 7 JD.com's Proprietary E-commerce

Jd.com started its e-commerce in 2004, and it became China's first comprehensive e-business listed in America (Nasdaq Index) in May 2014. In the past thirteen years, its turnover has increased 90,000 times, with a compound annual growth rate of over 150%, serving more than 900 million users and over 100,000 brand owners, having 120,000 full time employees and creating four million job opportunities.[22]

JD.com is committed to being a one-stop comprehensive shopping platform. Relying on e-business, finance and logistics, it has established a proprietary e-commerce development mode. Manufacturers cooperating with JD.com focus on the R&D and production of their products while JD.com is in charge of the circulation. JD.com has extended its service from transaction to storage, delivery, after-sale service and marketing, which form a complete service system of Internet retailing.

JD.com's house appliances' market share accounts for 62% of the national total in that field, and its 3C commodities' market share accounts for 50% of the total online market share. JD.com Fashion is the category with the fastest growth, accounting for more than 40% of new purchases. JD.com's One-million Convenience Stores Plan serves 6.8 million small and medium-sized stores nationwide and JD.com's Corporate Purchase market share accounts for over 45% of total of corporate purchase.

[22]Source: *JD.com's Financial Report of the Year 2016.*

3. **Online shopping has entered the mobile consumption era.**

In recent years, China's mobile shopping has witnessed rapid growth as a new growth point in online retailing. In 2015, mobile shopping rate was 54.8%, exceeding that of shopping on PCs for the first time. Mobile shopping is now a major online shopping channel. The year 2016 witnessed a sustaining growth of mobile consumption, and the number of the users of mobile APPs was 469 million, with the use rate increasing to 67.5% (See Fig. 3.21). The mobile shopping market scale increased to RMB3.3 trillion *yuan*, with a year-on-year growth of 57.9%, accounting for 70.2% of the total on-line transaction volume.[23] Mobile APPs have become the major battlefield for leading e-businesses' competition. Mature traditional e-businesses represented by Tmall and emerging mobile e-businesses represented by mogujie.com have all begun to provide mobile shopping services. Marketing is becoming increasingly diversified. 90% of e-businesses have launched their mobile APPs[24] while increasing their input in marketing at mobile terminals, speeding up the promotion of their service modes, introducing media elements to attract shoppers and exploring video guide for shopping. In 2016, the growth rate of the top 10 mobile APPs on China's online market was 27.1%, three times of that of shopping on PCs.[25]

3.3.2 Vertical E-Commerce Platforms Are Developing Fast

1. **Vertical B2C e-commerce platforms are witnessing fast development and providing refined products.**

Vertical B2C e-businesses aim at their segmented fields, trying to provide refined products and specialized services, and make their marketing socialized to meet the diversified and personalized demand of consumers. In the vertical e-commerce fields like cosmetics and fashion, there are vip.com (vipshop.com), jumei.com, mogujie.com and lefeng.com, among which vip.com has been listed with New York Stock Exchange and become the leader in its field. As for vertical e-commerce of foodstuff, womai.com, relying on the supply system of COFCO Corporation, and sfbest.com, relying on the fresh food logistics of sf-experss.com, have showcased their unique advantages in their trade. Some other vertical e-businesses have also been successful on their segmented markets, for example, jiuxian.com in wine e-commerce, redbaby.sunning.com in mother and baby product e-commerce, boqii.com in pet product e-commerce, dfmeibao.com in jewelry e-commerce, mbaobao.com in case and bag e-commerce and automotive aftermarket e-businesses like tuhu.cn, qccr.com and guazi.com.

---

[23]Source: iResearch, *Research Report on China's Mobile E-commerce (2017)*.

[24]CCID Think Tank, *Series of Blue Paper of China's Industry and Informatization (2016–2017)*.

[25]Source: iResearch, Research Report on China's Mobile E-commerce (2017).

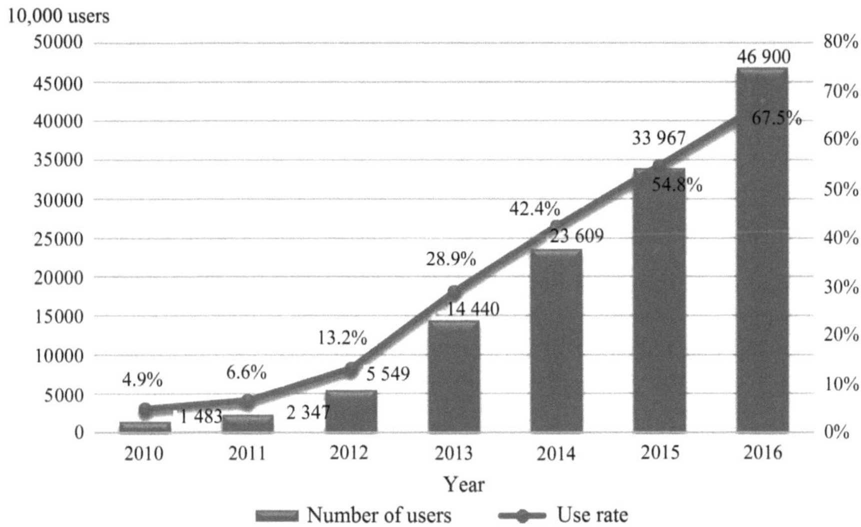

Fig. 3.21 China's mobile online shopping users and use rate (2010–2016) (*Source*: CNNIC, 39th Statistical Report on China Internet Development)

Column 8 Fashion Sales of Vip.com

As one of the leaders in vertical e-commerce, vip.com has formed the fashion sales mode of "selective brands, high discount and flash sale". Founded in Guangzhou in August 2008, the corporation was listed with New York Stock Exchange in March 2012. Its net income in 2016 was RMB56.59 billion *yuan*, with its annual order volume of 269.8 million and daily order volume of 800,000.[26]

Vip.com sells clothes, shoes, bags, cosmetics and household goods of 20,000 brands, including 2,200 exclusive cooperative brands. It has over 300 million members. As one of first e-businesses that entered the segmented market, it not only offers discounts and flagship brands, but also selective purchase guidance and financial and logistic service, having won the approval of consumers, especially stylish women consumers. In June 2017, vip.com upgraded its brands and changed its mission of "being a website dedicated to sales" to "being a website dedicated to sales of globally selected brands and quality goods". It has also launched innovation and upgrading in terms of APP, intelligence technology and logistics. From the third quarter of 2015 to the second quarter of 2017, it remained in the third place of its trade for eight

[26]Source: vip.com.

quarters in a row. It is now the world's biggest sales e-business and China's third largest e-commerce platform.[27]

2. **Vertical B2B platforms are expanding rapidly into manufacturing.**

Vertical e-commerce platforms provide opportunities for corporate e-commerce development. By making use of the Internet's efficient connection between supply and demand and reduction of information asymmetry, they help corporations to solve problems like excess production capacity, high circulation cost and insufficiency in effective supply. E-commerce of China's enterprises of iron and steel, petroleum and chemical, coal, nonferrous metals and automobile develops fast, and thus a number of vertical B2B e-commerce platforms have emerged with the transaction volume of over RMB10 billion *yuan* to even 100 billion *yuan*. It has become an important means of traditional industrial enterprise transformation to establish their vertical e-commerce platforms and get deeply involved in vertical platform systems. B2B e-businesses engaged in raw material transaction like youboy.com directly connects raw material suppliers with purchasers through its platform, which improves the transaction efficiency and reduces the cost of downstream purchasers; zhaogang.com engaged in iron and steel e-commerce simplifies the iron and steel transactions, and thus improves the efficiency of the supply chain; China Coal Solution, a coal supply chain business, provides service covering the whole supply chain from upstream to downstream through its bulk commodity (mainly coal) e-transaction platform yimei180.com.

3.3.3 Cross-Border E-Commerce Is Thriving

With China's opening up and improvement of consumption capacity in the country, cross-border e-commerce has become one of the important trends of e-commerce. Governments of all levels create all convenience for it. For instance, the State Council approved the establishment of 13 cross-border e-commerce demonstration zones in Hangzhou and Tianjin and other places in March 2015 and January 2016 to provide more replicable experience for cross-border e-commerce development; in 2016, Ministry of Finance, General Administration of Customs and State Administration of Taxation jointly issued the *Tax Policies on Cross-Border E-Commerce Retail Imports* and *List of Cross-border Online Retailing Imports* to promote regulated, mechanized and scale development of cross-border

[27]Source: *Report on China's B2C Market of Online Retailing*, released by Analysys from the third quarter of 2015 to the second quarter of 2017.

e-commerce. Meanwhile, large e-businesses are accelerating their cross-border e-commerce. For example, Alibaba Tmall's Double 11 in 2016 attracted 235 countries and regions to get involved. Supported by policies and driven by the market, cross-border e-commerce in China is developing fast. In 2016, the number of haitao.com users reached 41 million, with an increase rate of 78.3%, and the total transaction volume of the country's cross-border e-commerce (including retailing and B2B) amounted to RMB6.3 trillion *yuan*, with a year-on-year growth rate of 23.5%.[28] In the same year, the total cross-border e-commerce transaction volume of the above-mentioned 13 demonstration zones was over RMB160 billion *yuan*, the value having doubled.[29] In the future, with further promotion of the "Belt & Road Initiative", cross-border e-commerce will see a better prospect.

3.3.4 E-Commerce in Rural Areas and County Regions Is also Thriving

In recent years, China has launched policies supporting e-commerce in rural areas and county regions, covering pilot projects, innovation encouragement, channel expansion and financial service. Two-way circulation channels, namely, upgoing of agricultural produce and downgoing of consumer goods, have been set up. The number of e-businesses in county regions has grown from 10,000 to one million, and there have emerged e-commerce modes with regional characteristics, such as Suichang Mode and Shaji Mode. E-commerce is becoming a key means promoting synergetic development between urban and rural areas, transformation of county regional economy and targeted poverty alleviation. In 2016, the online retail volume in rural areas of China amounted to RMB894.54 billion *yuan*, accounting for 17.4% of the country's total. In the same year, there were 1,311 Taobao villages[30] with a year-on-year growth rate of 68%; 135 Taobao townships, with an increase of 90% in comparison with the number (71) in 2015, creating 840,000 jobs.[31]

---

[28]Source: Ministry of Commerce and General Administration of Customs (*Research Report on China's Cross-border E-commerce Market (2016–2017)*, edited by iResearch).

[29]Source: Wang Bingnan: Status Quo and Development Trend of E-commerce both at Home and Abroad: 29th *Special Lecture for the Standing Committee of the 12*th *National People's Congress (NPC)*.

[30]Taobao village refers to a village where the annual e-commerce turnover amounts to over RMB 10 million *yuan*.

[31]Source: AliResearch, *Report on E-commerce Development in China's Rural Areas (May 2017)*.

Column 9 Suichang Mode and Shaji Mode of China's E-commerce Development[32]

Suichang Mode: Suichang County, located in Lishui of Zhejiang Province, has formed the county region e-commerce mode "comprehensive e-commerce service providers + e-businesses + traditional industries", by promoting the e-commercialized development of local agricultural product processing with the support of policies. The core of the Suichang Mode is the localized e-commerce "service providers". They are Suichang E-E-businesses' Association established by Pan Dongming, who returned to his hometown in 2010 to start his business, Suichang Youth League Committee and some local businesses, and the "Online Store Service Center" affiliated to it. These organizations integrate the available goods sources, organize the distribution groups like local e-business people and centralize storage and delivery services. Besides, they have set up production and sale standards and unified data packages for farm and forestry products, and lowered the barriers in terms of technologies and funds, so that both producers upstream and sellers downstream can focus on their work. Suichang Mode is a great success in mountainous areas, setting a good example for more counties with weak e-commerce foundation, unknown brands and small-scale e-businesses, attracting leaders of different levels and scholars to visit the county to do some survey, and making Suichang one of the bases for survey and research on e-commerce in China's rural areas.

Shaji Mode: Shaji Mode originated from Dongfeng Village of Shaji Township, Xuzhou of Jiangsu Province. In Shaji, e-commerce does not rely on traditional industries, but has developed from "nothing". In 2006, the three e-businessmen led by Sun Han began to do e-business of simple furniture in Shaji. In that field, there is not so much difficulty in technology, fund and resource integration, and there is socialized division of labor among plank suppliers, mills, online stores and storage providers and packers. Besides, the furniture is sold through online stores owned by many farmers. In Shaji Mode, local farmers are dominators in information communication and resource integration. The mode has been replicated and hence it has led to the formation of a growing industrial cluster. E-commerce platforms represented by Taobao have become the important basic facilities for marketized public e-commerce, while agencies of logistics, telecommunication, power supply and credit as well as the government all create favorable environment for the development of the rural e-commerce. The success of Shaji Mode is a demonstration and inspiration for the middle and west regions with weak industrial and agricultural foundation.

[32]Source: AliResearch, *AStudy of Suichang Mode and A Survey Report on Shaji Mode.*

1. **E-commerce in rural areas (especially in county regions) have enriched the farmers' sources of income.**

Traditional agricultural production and sales are faced with dual difficulties, namely, drop of the price of agricultural products and high production cost, so e-commerce has become a powerful way of optimizing the agricultural product supply chain system. It helps supply and demand to connect with each other directly and traditional agricultural transformation to upgrade fast. Taobao, Tmall, JD.com and Suning are important platforms for sales of agricultural products, and SFBest and benlai.com are the key B2C platforms, which has led to the modern agricultural development mode of brand-building and refined operation. Origins of agricultural produce have begun to set up their own e-commerce platforms and micro-distribution platforms in addition to physical stores, and thus a diversified distribution system containing both online and offline stores has been established. In terms of supply, e-businesses are involved in organization and selection of supply of goods. Mode of direct purchase from production bases has been adopted at the platforms of Tmall agriculture and Tmall.HK, and regional products like those from Chengxian County of Gansu and Yangcheng Lake hairy crabs are sold at the platforms, which has promoted the commercialization and standardization of supply of goods. In terms of consumption, building of regional public brands and quality control traceability systems have enhanced the consumers' trust in the quality of agricultural products so they are willing to pay the premium of agricultural product brands. In 2016, the turnover of agricultural produce only on Alibaba platform amounted to over RMB100 billion *yuan*, with vegetables' turnover growing fastest, witnessing a year-on-year growth rate of 151%.[33] From 2012 to 2016, the total turnover of online retailing of agricultural produce in China increased from RMB 20 billion *yuan* to 158.9 billion *yuan*, with a growth of 8 times.[34]

2. **Downgoing consumer goods enrich the market in rural areas.**

Rapid development of online shopping in rural areas has resulted from the improvement of consumer goods' downgoing channels, which have tapped the potential of consumption on the rural market. Since 2014, Ministry of Commerce, Ministry of Finance and the State Council Leading Group Office of Poverty Alleviation and Development have jointly appropriated a fund of RMB8.4 billion *yuan* for the launch of comprehensive demonstration of the "E-business into Rural Areas" in 496 counties of 27 provinces (autonomous regions and municipalities directly under the Central Government). Thus three-level public service networks of rural e-business have been established, i.e. the levels of county, township and

[33]AliResearch: *From Revolution of Living Rooms to Revolution of Kitchens-White Paper of Alibaba's Agricultural Produce E-business (2016).*

[34]Source: Website of National Development and Reform Commission,*A New Era in Which Digital Economy Leads Economic Growth* (October 12th, 2017).

village, which have smoothed the channel for consumer goods to get into rural areas. Through Rural Taobao, 51ganjie.cn and rural e-post platform, businesses like JD.com, Suning, Taobao and China Post have set up 400,000 village e-business service stations in over 1,000 counties. Rural areas have become the main "battlefields" for e-businesses to compete with each other. In 2016, the online market scale in rural areas was RMB482.3 billion *yuan*, with a year-on-year growth rate of 36.6%. It was expected to reach RMB600 billion *yuan* in 2017.[35]

3.4 Equal Importance Is Attached to Internet Finance Development and Regulation

The thriving of Internet finance has brought about new momentum, solution and leapfrog development for real economy of financial service. The rapid rise of the third-party online payment, especially the mobile payment market, and the improvement of financial service quality and efficiency through Internet technology have promoted the rapid development of inclusive finance, online lending and online crowdfunding. To tackle the problems such as illegal fund-raising, false promise and information disclosure that have occurred with the development of Internet finance, the Chinese Government has formulated a series of rules and regulations represented by *Guiding Opinions on Promoting the Sound Development of Internet Finance*, carried out special rectification of Internet financial risks and improved the Internet finance supervision systems to promote the healthy development of Internet finance.

3.4.1 The Online Payment Service System Is Being Improved

As a traditional financial industry, payment and settlement has become one of the trades with the greatest vitality and brightest prospect by integrating itself with the Internet. In particular, with the popularization of mobile payment, online-payment has changed people's consumption habit and lifestyle.

1. **The online payment market is expanding year by year.**

To enhance the trust between supply and demand in e-commerce, Alipay serving Taobao.com was launched in 2004, which marked the beginning of third-party payment in China. After over 10 years of development, online payment has become an important part of the payment system in the country. By 2016, the market scale

[35]Source: China E-business Research Center, *Data Monitoring Report on China's Online Retailing Market (2016).*

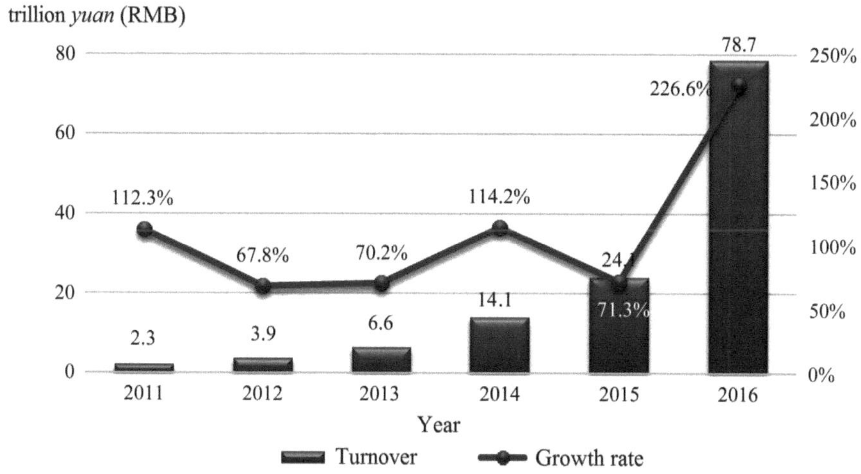

Fig. 3.22 China's third-party online payment market scale and growth rate (2011–2016) (*Source* iResearch, Supervision Report on China's Third-party Payment Market (2017))

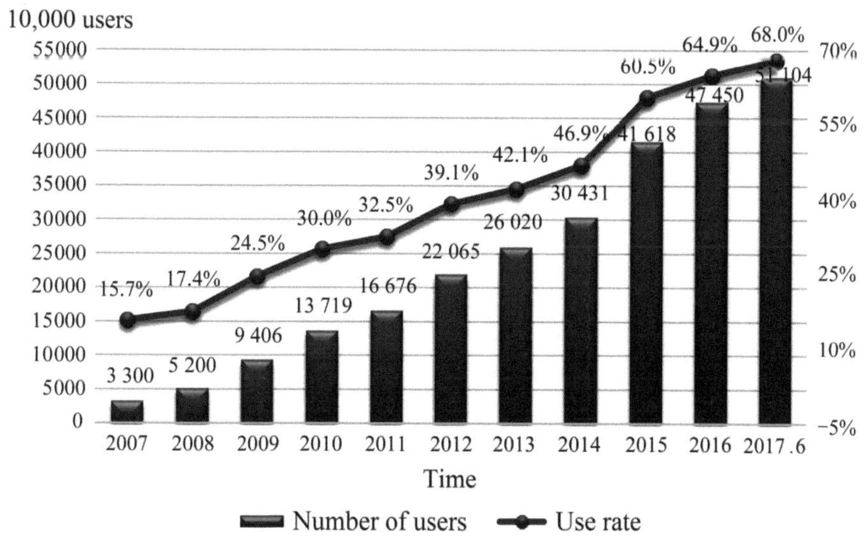

Fig. 3.23 Number of online payment users and use rate in China (2007–2017) (*Source* CNNIC, 39th and 40th Statistical Report on China Internet Development)

of third-party payment on PCs and mobile phones had reached RMB78.7 trillion *yuan* (See Fig. 3.22). By June 2017, the number of online payment users had reached 511 million (See Fig. 3.23) and the proportion of online payment by Internet users had increased from 64.9% to 68.0% in their total payment.

2. **The mobile payment market is expanding rapidly.**

In recent years, the mobile payment market is expanding at a dramatic speed. By June 2017, the number of mobile payment users had reached B502 million, with an annual growth rate of 31.2%, and the proportion of mobile payment by Internet users had increased to 69.4% in the total payment.[36] Alipay launched its mobile payment service in February 2008. 2.78 billion payments through it was done by 2013, and the amount was RMB 900 billion *yuan*, so Alipay became the largest mobile payment business in the world. In January 2014, WeChat's grabbing red packets became popular and over 8 million users got over 40 million red packets during the Spring Festival that year, which promoted the popularization of mobile payment APPs. WeChat TenPay and Alipay are the two giants in mobile payment service, with their total market share accounting for over 90% of the country's total. In 2016, China's mobile payment was over RMB81 trillion *yuan*, and the country is now the world's largest market of mobile payment. Chinese online payment businesses with Tencent and Alibaba as the leaders are accelerating their access to the international market. So far, Alipay and WeChat Pay have covered over 50 countries and regions and have over 200 million active cross-border users.[37] China's third-party mobile payment market shares in 2016 are shown in Fig. 3.24, and its third-party mobile payment amount and growth rate from 2011 to 2016 are shown in Fig. 3.25.

Payment businesses	Payment amount (RMB100 million yuan)
yqb.com	4207.7
UMF	2095.6
LianLian Pay	2079
JDPAY.com	2068.2
99Bill.com	1592.4
Yeepay.com	1439
China UMS	1242.7
Suning Pay	799.7

3. **Offline payment is penetrating the public life.**

After years of market cultivation and payment scenario development, China has formed rich offline payment scenarios, and the convenience created by mobile payment technologies like QR code and NFC (near field communication) have totally changed consumers' daily payment way and merchants' operation mode. Mobile payment has, to some extent, replaced the payment from physical wallets. Consumers have primarily formed the habit of using mobile payment tools in offline

[36]Source: CNNIC, *40th Statistical Report on China Internet Development*.

[37]Source: Sorted out of data published by Ant Finance and Tencent.

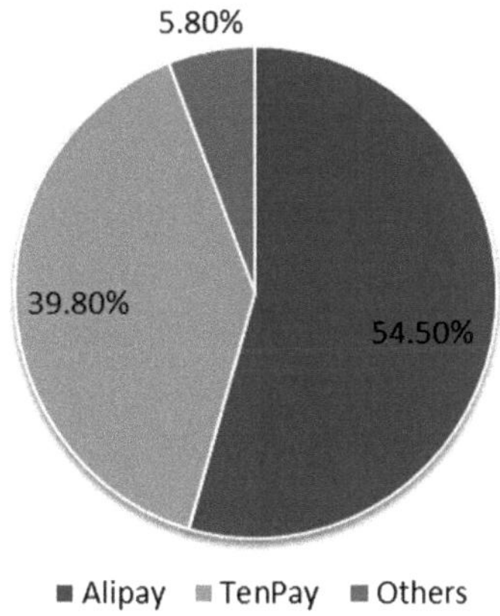

Fig. 3.24 China's third-party mobile payment market shares (2016) (*Source* ldhxcn.com, Report on the Third-party Mobile Payment in the Second and Third Quarters of 2017)

Payment businesses	Payment amount (RMB100 million yuan)
yqb.com	4207.7
UMF	2095.6
LianLian Pay	2079
JDPAY.com	2068.2
99Bill.com	1592.4
Yeepay.com	1439
China UMS	1242.7
Suning Pay	799.7

physical stores like restaurants, supermarkets and convenience stores, and such tools are penetrating third-tier and fourth-tier cities. The era of travelling with "wireless wallets" has quietly come. Online payment makes shopping convenient, reduces the sellers' bank card acquiring cost and cash management difficulties, and improves the experience of consumption service at the stores. By the end of 2016, mobile payment at physical stores had accounted for 50.3%. Offline payment APPs have great potential to penetrate different areas from cities to villages. Their use rate in fourth-tier and fifth-tier cities has reached 43.5% and 38.0% and that in rural areas, 31.7%.[38]

[38]Source: CNNIC, *39*th *Statistical Report on China Internet Development.*

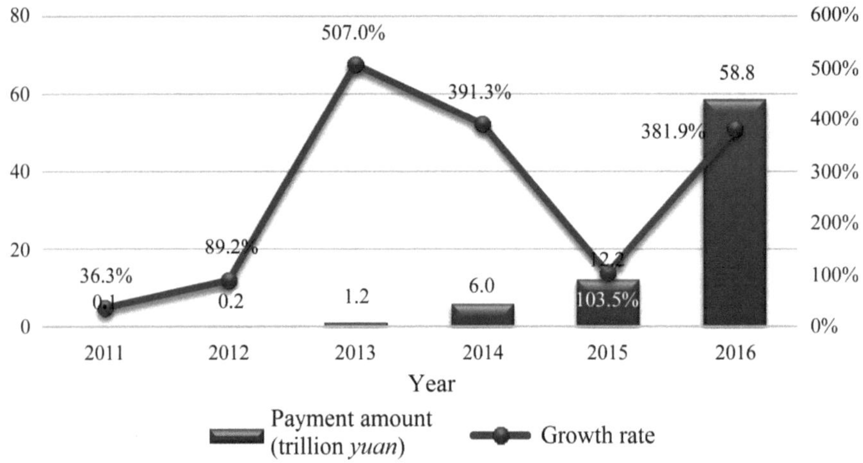

Fig. 3.25 China's third-party mobile payment amount and growth rate (2011–2016) (*Source* iResearch, Supervision Report on China's Third-party Mobile Payment Market (2017))

3.4.2 Financial Service Networking Is Being Rapidly Popularized

Internet financial service is not limited by time and space or any other barrier; instead, it can improve the users' experience, so it has totally changed the traditional financial service mode. In September 1999, China Merchants Bank became the first bank in China to launch the online banking service "all in one net". Approved by the People's Bank of China, it was the first to provide online personal banking service. In the past more than one decade, the online banking scale has been increasing, the products and services have been continuously innovated, and inclusive finance has emerged, which has sped up the Internet financial service in China.

1. **Financial institutions keep innovating their service mode.**

Traditional financial institutions like banks, stock exchanges, insurance companies, foundations and financial leasing institutions have sped up their integration with the Internet. Innovative Internet financial products and services based on cloud computing, big data and SNS have emerged and hence diversified the channels for investment and fund raising. In 2014, Internet banks saw their start, with the establishment of WeBank of Tencent and MYbank of Ant Financial being approved, and big data technology started to be used for credit rating. In 2015, Zhima Credit of Ant Financial first tested the personal credit consulting system, Weilidai of Tencent went online internally, JDbaitiao was upgraded to Baitiao+, and Baidu U-money and Tsinghua MBA began their cooperation in education credit.

2. **Inclusive finance helps small and micro businesses with their startup and innovation.**

"Inclusive finance" has been mentioned in the *Report on the Work of the Government* for three years and written into the *13*th *Five-year Plan*. Internet financial institutions have their natural advantages in perfecting credit data and reducing transaction cost. Through Internet technology, they can finish a large number of transactions, make the marginal cost of single transaction tend to be zero, and lower the access threshold. By relying on big data, they can dynamically adjust line of credit and eliminate the information asymmetry between the sides of credit. Since 2016, Internet corporations like Ant Financial, Baidu and JD have been cooperating with five State-owned banks in online and offline channels, covering part of areas that traditional financial service has failed to cover and improving the coverage rate, availability, pertinence and degree of satisfaction of financial service. "Inclusive finance" benefits micro, small and medium-sized businesses and "financially disadvantaged people", including farmers, low-income people poverty-stricken people, the disabled, and senior citizens, who can, through "inclusive finance", enjoy convenient and quick financial service at a reasonable price. In 2016, non-bank payment institutions like Alipay and Tencent TenPay had only 2.5% of the transaction volume in the total Internet payment, but the quantity of payments through them accounted for over 59%, and they served a large group of micro and small businesses. By May 2017, Ant Financial had provided over RMB one trillion *yuan* of credit loans for over six million micro and small businesses and individual entrepreneurs.

3.4.3 Online Lending and Crowdfunding Are Being Regulated

In recent years, online lending and crowdfunding have developed fast, which has given rise to some challenges and risks. To fight against illegal practice, guide businesses' running in accordance with regulations and guarantee the permanent, stable and healthy development of online lending and crowdfunding, a series of supervision policies have been launched and thus the online lending and crowdfunding are developing in a more regulated way.

1. **The online lending market has said farewell to wild growth.**

From 2013, China's online lending underwent its starting period and explosive growth, but by 2015, some online landing platforms had disappeared without any legal procedures, so online lending platform risk control and after-lending administration were to be strengthened. With the launching of regulations and policies like *Interim Measures for the Administration of the Business Activities of Online Lending Information Intermediary Institutions*, the online lending market has been regulated. The number of new online lending platforms reached its peak at the end

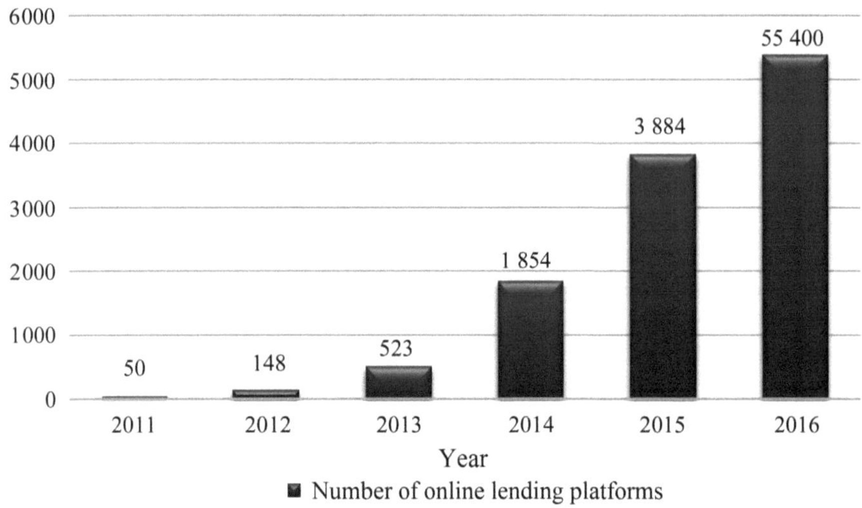

Fig. 3.26 Number and growth rate of online lending platforms in China (2011–2016) (*Source* ibaogao.com, Report on the In-depth Investigation and Investment Prospect Analysis of China's Online Lending Platform Market (2017–2022))

of 2014 and then began to drop. By the end of 2016, the number of online lending platforms had remained 5,400 (See Fig. 3.26), with the transaction volume of RMB652.621 billion *yuan*, 6.85 times that in the same period of 2014; the comprehensive interest rate of online lending had been 9.68%, with an annual decrease of 3.71%, which showed that the wave of high-interest-rate and high-risk speculative platforms and products had ebbed.

2. **Crowdfunding is being regulated.**

China's first crowdfunding website demohour.com was officially launched in July 2011, marking the beginning of the country's crowdfunding market. Since it can raise fund, intelligence and talents for micro and small businesses, crowdfunding has attracted a large group of entrepreneurs to engage in it. However, there exist many risks due to the variety of its fields, lack of laws or regulations on it, obscure supervision over it and chaotic operation within it. From the second half of 2015, the *Implementation Plan for Special Rectification on Risks in Internet Finance* and *Guiding Opinions on Promoting the Sound Development of Internet Finance* were released to reinforce the regulation of crowdfunding while encouraging and supporting its development. By the end of 2016, there were 752 crowdfunding platforms (See Fig. 3.27), among which 532 were running smoothly; there were 58,605 crowdfunding projects, among which 48,437 were successful, with the actually raised fund of RMB21.743 billion *yuan*.[39]

[39]Source: zhongchoujia.com, *Annual Report of China's Crowdfunding Development (2016).*

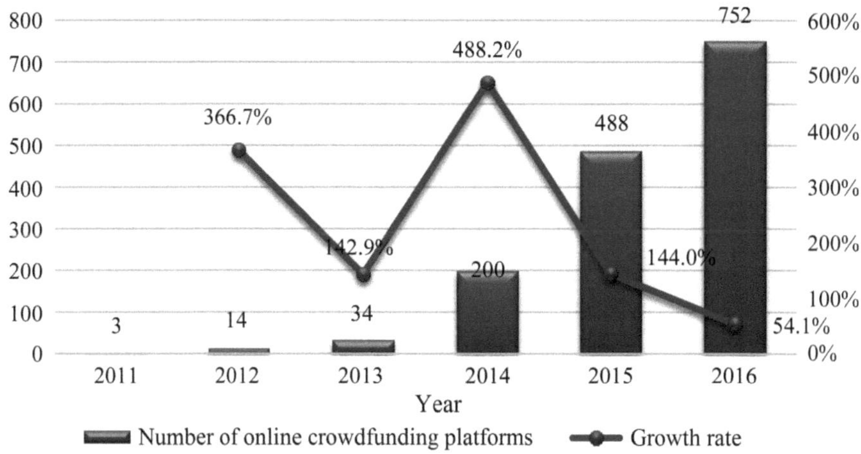

Fig. 3.27 Number of online crowdfunding platforms and their growth rate in China (2011–2016) (*Source* zhongchoujia.com, Annual Report of China's Crowdfunding Development (2016))

3.5 Sharing Economy Has Become a New Growth Point

Thanks to the rapid development of information technology and its innovative applications, sharing economy[40] has witnessed its blowout development. The year 2016 was called "the first year of China's sharing economy". Since then, bicycle sharing, house sharing, logistic sharing and financial sharing have all emerged and developed fast and become a totally new way of matching supply and demand, and optimizing resource allocation. In June 2016, *Self-discipline Pact on Internet Sharing Economy Service of China* was released and so was *Interim Measures for the Administration of Online Taxi Booking Operations and Services* in July of the same year, which shows that the policy guarantee, credit environment and industrial self-discipline are all being regulated for the development of sharing economy.

3.5.1 The Market Scale of Sharing Economy Is Witnessing Its Explosive Growth

At present, sharing economy of China is centered in public consumption and service, especially in transportation, housing and medical care (See Table 3.1). In transportation, Didi Chuxing and Uber develop fast. By June of 2017, the number of online taxi-hailing service users had reached 278 million, with a growth rate of

[40]Sharing Economy refers to the gathering, reuse and match between supply and demand of massive, scattered and idle resources through platforms in collaboration. It is a new economic and industrial form suggesting innovation of economic and social values.

23.7%.[41] In bicycle sharing, 20 million sharing bicycles were launched in 2017 and the number of the users had reached 106 million.[42] As for housing, Airbnb, tujia.com and xiaozhu.com had become the leading platforms. In 2016, the transaction volume of house sharing was RMB24.3 billion *yuan*, an increase of 131% in comparison with that in 2015, and there were 1.9 million houses for sharing and over 35 million participants. In medical care sharing, the transaction volume in 2016 was RMB 15.5 billion *yuan* and approximately 2.56 million people provided medical care service for 200 million people. In knowledge payment, there were such platforms as Zhihu, Guokr.com, ximalaya.com, igetget.com and lizhiweike.-com in 2016 and they had 50 million users. The year 2016 saw the sharing economy turnover of about RMB3.45 trillion *yuan,*[43] with a year-on-year growth of 103%. It is expected that the scale of sharing economy in China will account for over 10% of the country's total GDP in 2020. Sharing economy will probably become a subversive economic force in the coming decade. It is estimated that the global market value of sharing economy will be 335 billion US dollars[44] in 2025.

3.5.2 Entrepreneurship and Innovation Have Been Boosted

Sharing economy has led to the emergence of new patterns in transportation, smart APPs and household life. It has become the model of market innovation and cross-field entrepreneurship, lowering the threshold for innovation and entrepreneurship, reducing cost, increasing the speed and attracting more participants. So far, eight businesses in sharing economy have gone public within 66 months since they were founded, while it usually takes other Internet businesses 84 months to succeed in going public.[45] Among the 186 "unicorn" companies in the world, there are 42 Chinese ones, among which 15 are sharing economy ones, accounting for 35.7%.[46] The growth of startup and innovation businesses in sharing economy has also attracted the attention from the capital market. In 2016, RMB171 billion *yuan* was raised as fund for the sharing economy businesses.[47]

---

[41]Source: CNNIC, *40*th *Statistical Report on China Internet Development.*

[42]Source: Data from Chinese Cycling Association.

[43]Source: Sharing Economy Studies Center of State Information Center & Sharing Economy Working Committee of Internet Society of China, *Report on China's Sharing Economy Development (2017).*

[44]Source: Predicted data from PWC.

[45]Source: Tencent Research Institute, Report on Sharing Economy (2017), May 27th, 2017.

[46]Source: CBInsight data analysis based on *Report on China's Sharing Economy Development (2017).*

[47]Source: Sharing Economy Studies Center of State Information Center & Sharing Economy Working committee of Internet Society of China, *Report on China's Sharing Economy Development (2017).*

Table 3.1 Market scale of key areas of sharing economy in China

Area	Turnover in 2015 (100 million yuan)	Turnover in 2016 (100 million yuan)	Growth rate(%)
Knowledge and skills	200	610	205
House rent	105	243	131
Transportation	1, 000	2, 038	104
Daily service	3, 603	7, 233	101
Production capacity	2,000	3, 380	69
Medical care	70	155	121
Capital	10, 000	20, 863	109
Total	16, 978	34, 522	103

(*Source* Sharing Economy Studies Center of State Information Center and Sharing Economy Working Committee of Internet Society of China, *Report on China's Sharing Economy Development (2017)*)

Column 10 Bicycle Sharing

Bicycle sharing refers to bicycle renting at campuses, subway stations, bus stops, neighborhoods, business districts and public service areas. It aims to make full use of public passages. It is a new type of sharing economy based on periodic lease. In 2016, "bicycle sharing" became a nationwide popular concept and the number of uses of the bicycles reached 1.886 million.[48] According to statistics, through fierce competition, there are now over 20 businesses in the field of bicycle sharing. Ofo founded in 2014 and Mobike founded in 2015 are the representatives who cooperate with Ant Financial and WeChat Wallet respectively to occupy the domestic market. In 2017, they launched their business even into Singapore, the United States, Australia and Japan.

Compared with public bicycles locked to certain facilities, shared bicycles have brought much more convenience for urban residents since they are not locked to certain facilities but can be used much more freely. On the other hand, some of these bicycles are placed disorderly, or even on fire exit passages, which may result in safety hazards and hard management of urban space. Therefore, on May 7th, 2017, Bicycle Sharing Specialized Committee of Chinese Cycling Association was founded. It will participate in the formulation of *Bicycle Sharing Group Standards*. Hangzhou, Shanghai, Beijing and Zhengzhou municipalities have released their proposals on promoting the regulated development of bicycle sharing. On August 3rd, 2017, ten ministries including Ministry of Transport announced the *Guiding Opinions on Encouraging and Regulating the Development of Internet Bicycle Rental*, which proposes that bicycle sharing platforms should improve the online and

[48]Source: Monitoring data from China E-business Research Center.

offline service capacities, and clarifies the measures on regulating bicycle placement and setting up electronic fences.

3.5.3 A Large Number of Employment Opportunities Have Been Created by Sharing Economy

The fast growth of sharing economy has created a large number of flexible employment opportunities. In 2016, over 600 million people got involved in sharing economy, and among them, about 60 million were service providers, 10 million more than the number in 2015. A report released by Didi Chuxing shows that in 2016 the platform provided 17.5 million flexible jobs for society and that 2.384 million employees came from over capacity-cutting industries. In the field of house sharing, platforms like xiaozhu.com, Tujia and zhubaijia.com have directly or indirectly offered jobs to two million people. It is anticipated that by 2020, there will have been over 100 million sharing economy service providers. The social organization pattern "corporation + employees" will be replaced by "platforms + individuals".

Column 11 Didi Chuxing
Xiaoju Technology Co., Ltd. was founded in Beijing in July 2012 and Didi Dache APP was launched in September of the same year. In February 2015, Kuaidi Dache and Didi Dache announced that they would be merged strategically. The merger was rebranded into Didi Chuxing in September 2015. In October 2015, Didi Kuaidi Premier Platform received China's first internet car-booking license from the Shanghai Municipal Transportation Commission.

Today Didi Chuxing provides Didi Taxi, Didi Premier, Didi Express, Didi Hitch and Didi Bus, serving nearly 300 million users in over 400 cities. In the year of 2015, Didi announced that the platform fulfilled 1,430 million orders, a number higher than the total of Uber in six years, which was 1,000 million. So far, Didi has become the world's largest one-stop transportation platform and the world's second largest online transaction platform, only next to Taobao. It is expected that the platform will, based on electrified, automotive and sharing transportation, launch "Transportation 3.0", which will redefine transportation. Super hitch, smart traffic lights and smart drive will help to completely solve the problems such as traffic jam, traffic accidents, exhaust pollution and difficult parking.[49] Therefore, Didi will commit itself to

[49]People.cn-IT Channel, Liu Qing, President of Didi Chuxing: *The Future Transportation Will "Say Farewell to" Traffic Jam and Accidents*, June 27th, 2017, http://it.people.com.cn/n1/2017/0627/c1009-29366714.html.

following China's Internet innovation strategy through its practice in sharing economy. It will cooperate and compensate with different communities and partners and apply big data-driven deep learning technology to tackle the challenges in transportation and environmental protection.

3.6 "Internet+" Is Transforming and Upgrading Traditional Industries

In July 2015, the State Council issued the *Guiding Opinions on Vigorously Advancing the "Internet Plus" Action* to boost the deep integration between the Internet and primary, secondary and tertiary industries and thus push the Chinese industries to the mid and high end.

3.6.1 "Internet+ Service" Has Activated Innovation and Entrepreneurship

1. **Internet facilitates the steady growth of daily life service.**

As for "Internet+ transportation", Internet has led to the occurrence of a new system of comprehensive transportation services. ETC, e-card and e-ticketing all make transportation convenient. ETC serves nearly 40 million users.[50] The navigation electronic map provides location service-based intelligent information. As for "Internet + tourism", the online tour booking market in China has seen rapid growth. By the end of 2016, the number of the users had reached nearly 300 million, with a use rate of over 40% (See Fig. 3.28). As for "Internet+ education", the online education market scale in China reached RMB156.02 billion *yuan* in 2016, with a year-on-year increase of 27.3%. It is anticipated to maintain the growth rate of 20% in the coming years.[51] Besides, new technologies facilitate the O2O interaction and integration, and deep mining based on user data enhances the match between service and demand. O2O services are becoming more diversified in online taxi-hailing, catering takeout, household service, beauty and health care, education and training, and vehicle maintenance.

2. **Internet facilitates the transformation and upgrading of logistic service.**

The development of online retailing has led to the explosive growth of express delivery. From 2010 to 2016, the number of the expresses delivered increased from

[50]Source: Data released by Ministry of Transport in September 2016.

[51]Source: iResearch, *Research Report on Online Education Platforms of China (2017).*

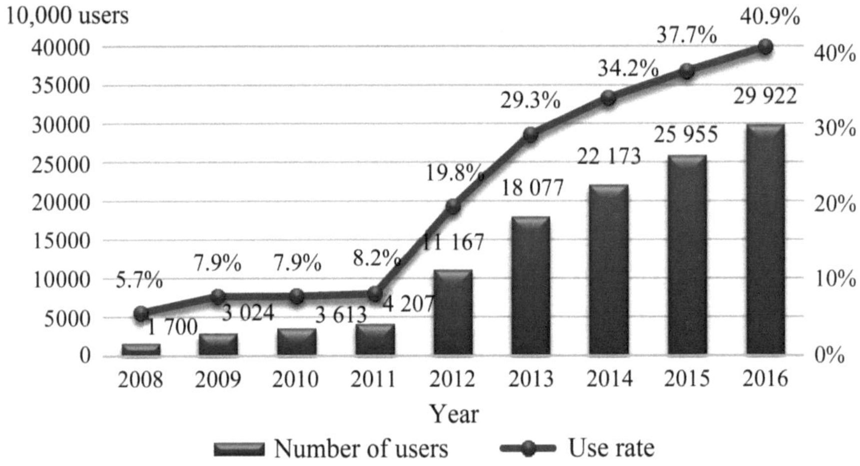

Fig. 3.28 Number of users and use rate of tour booking in China (2008–2016) (*Source* CNNIC, Statistical Report on China Internet Development)

2.34 billion to 31.28 billion, with the revenue rising from RMB57.46 billion to RMB397.44 billion. 60% of express delivery resulted from online shopping.[52] Internet helps to gather the demands and then deliver the goods at one time, which boosts the intensification and linearization of logistics, which, in turn, helps to reduce the cost of logistics and storage of the whole society and thus remove the bottleneck of high logistic cost of China. Meanwhile, Internet forces the logistic industry to transform and upgrade itself. Leading logistic businesses have begun to enhance the construction of logistic networks and their cooperation with e-businesses. They provide high-quality services like "next-day delivery to the destination in a different city", "same-day delivery to the destination in the same city" and "more than one delivery to one user within one day". By providing real-time parcel tracking service, and improving the insurance system and articles of liability for compensation, they keep improving their logistic service.

3.6.2 "Internet+ Manufacturing" Has Boosted Flexible Productivity

In May 2015, the State Council issued *Made in China (2025)* to comprehensively implement the strategy of strengthening the country through manufacturing. In May 2016, it issued the *Guiding Opinions on Enhancing the Integrated Development of the Manufacturing Industry and the Internet*, which proposes to cultivate and

[52]Source: State Post Bureau, *Index Report on China's Express Development.*

develop new products, modes and industries of the integration between manufacturing and Internet, to improve the quality, efficiency of manufacturing and to transform and upgrade manufacturing through Internet. Internet technologies are used in production, which helps to improve manufacturers' competitiveness in collaborative R&D, lean management, risk control, supply chain collaboration and rapid market response. In 2016, the popularization rate of digitalized R&D tools and networking rate of digitalized manufacturing equipment of industrial businesses in China respectively reached 61.8% and 38.2%[53]; as for the manufacturers that have passed the management system standard of "integration of informatization and industrialization", the R&D cycle of their new products was shortened by 32.6%, the design change rate was reduced by 25.7%, the production plan completion rate increased by 3.7%, the users' complaints dropped by 48.6%, and the comprehensive profit was obviously improved.[54] "Internet+" pushes flexible manufacturing, personalized customization and service-oriented manufacturing to rise at a high speed. Manufacturers of different fields such as Redcollar, Haier, Wision and Xiaomi Technology have founded their C2 M factories and new online direct selling patterns like M2C and F2C. ShannGu, XCMG and DEC are developing life-cycle management for their Internet-based products, remote operation and maintenance and financial leasing. They are also transferring local process optimization to whole process reengineering and single business application to multi-business synthesis. The cloud manufacturing system has been formed fast with cloud design, cloud production, cloud management and cloud test as the cores. CASIC, CASICloud, Beijing Cloud Chain Industry SaaS Cloud Platform and XCMG-Cloud have been launched online. The cloud manufacturing platform for household industry in Foshan City has attracted over 30,000 businesses to join it.[55] The rapid development of industrial Internet provides key support for intelligent manufacturing. In 2016, the market scale of China's industrial Internet reached RMB189.6 billion *yuan*, with a year-on-year growth rate of 27.33%. According to the CCID, it is anticipated that the compound annual growth rate will remain around 25%. In the coming two decades, the development of China's industrial Internet will help the country's GDP to increase by US$ 3 trillion.[56]

[53]Source: Miao Wei (Industry and Information Technology Minister), *Work Report at the Seventh Meeting of the Standing Committee of the 12*th *National People's Congress.*

[54]Source: *Success Has Been Achieved in the Management System Standard of "Integration of Informatization and Industrialization*, http://www.miit.gov.cn/n1146290/n1146402/n1146440/c5512179/content.html, March 7th, 2017.

[55]Source: Official data from "China's online manufacturing" on the public cloud service platform in manufacturing of Foshan City.

[56]Source: Miao Wei (Industry and Information Technology Minister), *Work Report at the Seventh Meeting of the Standing Committee of the 12*th *National People's Congress.*

3.6.3 "Internet+ Agriculture" Is Revitalizing China's Agriculture

In recent years, a series of national policies on "Internet+ agriculture" have been launched, pushing the Internet to penetrate R&D, market circulation and financial support of agriculture, facilitating the improvement of agricultural product consumption structure and added value, and helping the traditional agriculture to be revitalized. As for R&D, the Internet helps agricultural technology consulting to be popularized. Large-scale, specialized and intensified production has been realized through the combination of new-generation information technology and traditional agricultural management technology and thus the informatization and intelligentization of agricultural production has been improved. So far, Ministry of Agriculture has launched pilot and demonstration projects of IoT in agriculture in eight provinces, having introduced 426 technologies, products and applications, and established National Agricultural Data Center, National Agricultural Scientific and Technical Data Sub-center and 32 provincial agricultural data centers.[57] As for market circulation, Internet businesses have begun to input their technology and fund into e-commercial markets in rural areas. Networking keeps improving in socialized services of agriculture, such as market consulting, marketing design and targeted logistics. Cuncunle.com, the biggest Internet platform in rural areas of China, is dedicated to Internet entrepreneurship, integrated marketing and material circulation in rural areas. It has set up an online portal for each of more than 600,000 villages nationwide, with more than 12 million registered members.[58] In agricultural finance, Internet platforms provide services for every link of the agricultural chain, covering credit investigation, payment, money management, loaning and financial leasing. By April 2018, the accumulative turnover of only "Nongjinquan", a leading financial service for rural areas, had amounted to over RMB14.6 billion *yuan*..[59] Moreover, the construction of "Internet+ quality safety" platforms have made the safety of the whole chain of agricultural product processing traceable, and "Internet+ leisure agriculture" promotes the development of agricultural tourism and agricultural product experiential consumption, contributing to the diversification of modern agricultural profit channels and opportunities, and the enlargement of agricultural terminal market consumer groups.

[57]Source:"Internet+" modern agriculture witnesses a new wave of agricultural informatization, http://www.moa.gov.cn/xw/zwdt/201612/t20161216_5408744.htm, January 18, 2017.

[58]Source: http://www.cuncunle.com/company_profile.

[59]Source: Data released by Nongjinquan. https://www.lcfarm.com/info_platformdata.html.

Chapter 4
E-government Makes Governmental Service More Efficient and Transparent

Chinese Academy of Cyberspace Studies

General Secretary Xi Jinping points out that information is a key foundation for national governance and that we should boost the modernization of the national governance system and governance capacity through informatization. Since 1980s, China's e-government has experienced three periods. The fist was the starting period, when the OA project for CPC organizations and governmental agencies was the focus and other government informatization projects such as Government Networking Project were also launched. The second was comprehensive launch period, when the CPC Central Committee and the State Council released a series of policies and documents, proposing that e-government construction should be a focus of informatization of China in the following years and encouraging the government to take the lead in promoting the informatization of national economy and social development. Thus e-government construction was launched comprehensively. The third is today's innovation and transformation period, when China's e-government is witnessing the transformation of its construction goal, approach and mode. In this period, China tries to innovate the social governance approach, break information barriers, improve the people's living, better know about the social situation, smooth communication channels and support scientific decision making through informatization.

Chinese Academy of Cyberspace Studies
Beijing, China

Chinese Academy of Cyberspace Studies (ed.), *China Internet Development Report 2017*,
https://doi.org/10.1007/978-3-662-57521-5_4

4.1 The Security System Supported by E-government is Being Improved

4.1.1 The E-government Network is Basically Completed

1. ***Jin* Projects (Golden Projects) lay a foundation for the e-government network framework.**

At the end of 1993, China officially launched the "three *jin* projects" (three golden projects), namely, *jinqiao* (golden bridge), *jinguan* (golden customs) *and jinka* (golden card), which have promoted the rapid development of the national economy. In August 2002, General Office of the CPC Central Committee and General Office of the State Council released *Guiding Opinions of the Leading Group of the National Informatization on National E-government Construction*, in which "two networks, four databases, twelve *jin* systems" were proposed, covering all areas of e-government. Thus the primary framework of China's construction was thus formed. Through the *Jin* Projects (Golden Projects), the Chinese Government has accelerated the construction of informatization infrastructure and improved the network-based management of all fields. Thanks to years of development, a number of key national information application systems such as *jinguan* (golden customs), *jinshui* (golden taxation) and *jindun* (golden shield/police affairs) can be called advanced systems in the world, having dramatically improved the governmental service efficiency. By the end of 2016, social security "e-card" had been issued by Ministry of Human Resources and Social Security to 972 million people, 70% of the total population. The nationally unified social security card had been issued in 31 provinces, autonomous regions, municipalities directly under the Central Government and Xinjiang Production and Construction Corps, covering 96.6% of prefectures and municipalities; an average of 80 service directories had been launched throughout the country[1]; immediate settlement of medical care expenses, cross-regional card use, pension drawing and personal on-line payment of social security fund have been gradually launched and become popular in society.

Column 12 *Jin* Projects

Three *Jin* Projects includes *jinqiao* project (golden bridge), *jinka* project (golden card) and *jinguan* (golden customs). *Jinqiao* project means that a public economic information network is established throughout the country to connect with the special networks of all ministries. *Jinguan* project is a network of the information system of foreign trade businesses throughout the country. In this project, EDI technology is popularized, no paper used for

---

[1]Source: *Big Data Report on the National Social Security Cards* released on http://www.mohrss.gov.cn.

trade. *Jinka* project is a currency electronization project with the use of Information Card and Cash Card as the aim.

"Two networks, four databases, twelve *jin* systems" refers to the construction and integration of a unified government network made up of the internal and external administration networks; planning and development of population information base, legal person information base, natural resource and spatial geography base and macro-economy database; construction and improvement of the governmental office resource system and key work systems like *jinguan*, *jinshui*, financial supervision (including *jinka*), macro-economic management, *jincai* (golden finance), *jindun* (golden shield/police affairs), *jinshen* (golden auditing), social security, *jinnong* (golden agriculture), *jinzhi* (golden quality) and *jinshui* (golden hydraulic engineering).

2. **Basic supporting capacity of the e-government network has been enhanced.**

In September 2007, the Central Government backbone transmission network of the national e-government networks begun to provide service, marking that the basic carrier supporting the construction and operation of internal and external government networks was primarily established and that a unified national e-government network framework was basically formed. The network is a reliable, resourceful, regulated and professional platform for all departments and regions. After years of development, the construction of internal e-government network platforms and top-level interconnectivity is progressing steadily, and the network is playing an increasingly obvious supporting and ensuring role. By the end of 2016, 32 provincial internal government network platforms and 15 sub-provincial ones had been completed. The internal government network platforms of municipal or prefectural level and below are being constructed on the principle of "safety guarantee, and construction in accordance with the demand". Online governmental service platforms in Zhejiang, Guangdong and Guizhou provinces had been able to cover the four levels of province, municipality, county and township. Those of eight provinces had been able to cover three levels, namely, the provincial, the municipal and the county.[2] Meanwhile, the switch rate of external government networks keeps increasing. By the end of June of 2016, about 147,000 governmental organizations had been successfully connected to the external government networks of different levels, with over 1,844,000 access terminals. Altogether, 130 central governmental organizations had successfully connected to the Central Government-level external

[2]Source: *Report on China's E-government Development* (2017)*: "Internet+" helps to innovate the governmental service mode.*

network platform while the rate of access to provincial ones had reached 100%; that of access to municipal or prefectural ones, 94.6%; that of access to county ones or below, 89.5% and 34.1% respectively.[3]

4.1.2 The E-Government Governance Systems and Regulations Are Being Improved

Since the founding of the Central Leading Group for Cyberspace Affairs, planning and coordination for e-government development have been strengthened and the initial "top leadership project" has been transformed into the legalized, institutionalized and specialized e-government governance system. The leading group for cyberspace affairs has been established in all provinces, some of which, for example, Shaanxi and Gansu, were the first to have taken the leading group as the planner and coordinator of local e-government, which, in turn, has boosted the development of local e-government. In recent years, the Central Government has launched some policies and documents to clarify the construction goal, construction approach and system mode of e-government. In 2016, "Internet + e-government" was written into the *Report on the Work of the Government* for the first time, producing profound influence on the transformation of the government function and the economic and social development and bringing about new mode of China's e-government development. *National Informatization Planning for the Thirteenth-five-year Plan Period* has listed "Internet + e-government" as one of the 12 priorities. In September 2016, the State Council issued the *Guiding Opinions on Accelerating the Promotion of the "Internet + Government Services"* to improve the comprehensiveness of online service, information sharing, probability of service application handling and convenience for businesses and individuals. To promote the progress of e-government, the State Council formulated and issued more documents, including *Guidelines on further Enhancing Governmental Departments' Information Sharing Construction Management*, *Interim Measures for the Administration of Sharing of Government Information Resources*, *Implementation Plan for Governmental Information System Integration and Sharing*, and *Opinions on Comprehensively Advancing the Work of Open Government*, pertinent to coordination, information sharing, resource optimization and integration and management and operation mechanism, to solve the problems hindering China's governmental informatization, such as fragmentation of information and isolated administration. By the end of 2016, over 30 ministries, including General Office of the CPC Central Committee, General Office of the State Council and National Development and Reform Commission, had launched nearly 100 e-government

[3]Informatization Research Department of State Information Center: "Present Status and Prospect during the 13th Five-year-plan Period of China's E-government", Informatization Research, Vol. 8, 2017, http://www.sic.gov.cn/News/258/8150.htm.

plans and measures, about 43% concerning the operation system, 26% concerning security, 13.9% concerning web portals and 12% concerning standards and regulations.[4] They create a favorable environment for the healthy, coordinated and sustainable development of e-government.

4.2 E-government Service Is Being Improved

4.2.1 Governmental Websites Are Playing a More Prominent Role in Public Service

China's "Government Networking Project" was officially launched in 1999. Since then, all regions and departments have been active in promoting the construction of governmental websites to provide the public with online service. Today, 96% of China's governmental websites provide online service, which covers the demand of both individuals and businesses, with the full-process service rate of 73.3%.[5] Governmental websites gather information and service resources of all departments to provide cross-section and cross-field online services, covering daily life, medical care, social security, housing and transportation. In 2016, the Chinese Government issued *Guiding Opinions on Accelerating the Promotion of the "Internet* + Government Services", proposing that by the end of 2017, an integrated online e-government service platform would be built to make all government services open to the public. By the end of 2020, an Internet + governmental service system of integrated linkage, collaboration among departments, coordination by provinces and service within one network will have been established to improve the intelligent level of governmental service. Online services for individuals include family planning, residence registration, education, culture, employment, military service, marriage, taxation, social security, transportation, medical care, postal service, exit from and entry into China, and legal assistance; services for businesses cover registration for startup of businesses, examination and review, investment and taxation, labor security and other kinds of services closely related to businesses.

Column 13 Gov.cn Launched the Column "I have a word to say to the Premier".

In 2014, the Chinese Government Network (gov.cn) launched the Column "I have a word to say to the Premier" to gather the public opinions. It is a new governmental service platform, gaining popularity and praises among Internet users. According to statistics, the number of messages rose from 96,774 in 2014 to 404,090 in 2016, with a year-on-year growth rate of 96.5%. The

[4]Source: *Report on China's E-government Development* (2017).

[5]Source: CNNIC.

government personnel treat every message seriously, collecting all the opinions and suggestions and studying them. Some good ones are reported to the Premier. 2,071 messages of more than 400,000 in 2016 were submitted to the State Council leaders and the drafting group of the *Report on the Work of the Government*. Of the 2,071 messages, 101 ones were consistent with the *Report*. Therefore, the column has become a bridge between Chinese Internet users and the Government. For example, on July 15th, 2016, Hongyan, a 40-year-old from Beijing, left a message which said, "Please do not let those on business to apply for an invoice with the license of his business with him, but simplify the procedures for business tax being replaced by value-added tax." Other 1,974 Internet users left similar messages concerning "business tax being replaced by valued-added tax". On December 29th, 2016, Xiao Jieming, then Minister of Finance, said that by the end of November that year, a reduction of RMB338.7 billion *yuan* had been done since the taxation reform (business tax being replaced by VAT) was overall promoted. On December 27th, Wang Jun, Director-General of Sate Administration of Taxation, said that the reduction would be even higher in 2017.

Column 14 "One-window Service" in Guangzhou and "One-door Service" in Foshan

In March 2016, the General Office of the People's Government of Guangdong Province issued the *Implementation Plan on Promoting One-door and One-website Service throughout the Province*, which proposes that all the content directories of public services should be sorted and made known to the public, that certifications and procedures should be cut off, and that simplification, and optimization and service innovation should be promoted. According to the *Plan*, unified standards should be formulated for licensing and services, the province and its municipalities should provide one-website services and a platform for online office hall should be set up to connect seamlessly with the departmental examination and approval system. The service model should be established of comprehensive acceptance at the reception, classified examination and approval at the back office, and licenses or other documents being issued out of the same window. One-door and one-website service model is being tested on administrative licensing service for legal persons, and one-door service at the grass-root level is being promoted concerning licensing and service for natural persons.

4.2.2 New Media of E-government Are Making Social Condition and Public Opinion Communication Channels More Convenient and Diversified

1. **The number of new media of e-government has seen a dramatic growth.**

Following the development of the mobile Internet and new media, the Chinese Government takes Weibo, WeChat, and APPs as important channels to communicate with and serve the public, promoting the rapid development of new media for e-government affairs. In 2011, Cyber Administration of China held a symposium on serving society through Weibo. In February 2015, it held an experience sharing meeting on the construction and development of new media for government affairs, calling on the CPC and governmental organizations in different regions, enterprises, public institutions and mass organizations to use instant communication tools for their information service. Weibo, WeChat and APP were taken as the new media for e-government for the first time at the meeting. They have seen rapid growth ever since. By June 2017, the number of governmental Weibo accounts verified by sina.com had amounted to 170,000; that of WeChat official accounts, 514,000; and that of Toutiao official accounts, 34,083.[6] Governmental Weibo, WeChat, APP and Toutiao accounts play an important role in publishing authoritative information, carrying out government-people interaction and guiding online public opinions. According to the analysis by people.cn, among the over 600 public opinion cases from January to November 2016, the governmental response rate amounted to 87%, and 57% of the accidents received the government's first response within 24 h; 73% within 48 h; and 41% through the governmental new media.[7] The general layout of governmental Weibo accounts is shown in Fig. 4.1.

2. **Mobile government channels launched by new government media.**

The development of new government media has sped up the spread of governmental information, making the governmental services more diversified, people-friendly, interactive and active. Governmental agencies have enhanced their cooperation in new government media construction to promote the popularization and convenience of new government media. By the end of 2016, 362 cities of the country had opened online urban service to the public through the WeChat platform, benefiting over 400 million users, and covering over 50 services such as social security, transportation, taxation, education, medical care and payment for water, power and gas.[8] Alipay supports the daily life service in 27 provinces,[9]

[6]Source: CNNIC, *39th Statistical Report on China Internet Development*.

[7]*Assessment Report on the Index of Response to Public Opinions of China* (2016), New Media Think Tank of people.com.cn.

[8]Source: Tencent, *White Paper of the Development of "Internet + Human Resources and Social Security"*.

[9]Source: Alibaba.

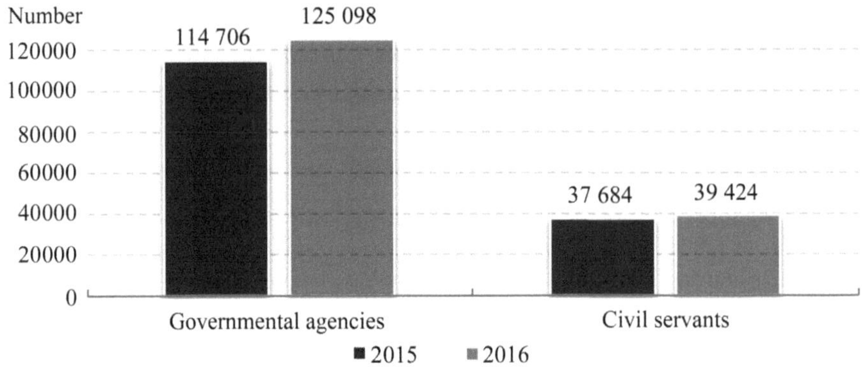

Fig. 4.1 General layout of government Weibo accounts (*Source* Sina Weibo)

Alipay supports the daily life service in 27 provinces,[9] and through the government affairs supermarket on the Alibaba Cloud Computing Platform, urban residents can obtain governmental services in the way they shop on Taobao. For example, thanks to this platform, three levels of government (namely, the provincial, the municipal and the county governments) can handle over 60,000 applications for examination and approval on the same website and the payment can be done online.[10]

4.3 Information Resource Sharing Is Promoting Governmental Service Modernization

The Chinese Government always attaches importance to and promotes information resource sharing. It takes governmental data development and utilization as an important means of releasing the digital bonus and boosting the economic development. It tries to seek out new practice of governmental data resource development and utilization and new modes of information resource sharing while facilitating open and transparent government, and promoting fairness and justice as well as inclusive development of e-government.

[9] Source: Alibaba.

[10] Source: Comprehensive reports on zjzwfw.gov.cn by huanqiu.com and dahe.cn.

4.3.1 Information Publication and Data Openness Are Facilitating the Construction of Transparent and Efficient Government

1. **Governmental information publication is being accelerated.**

In May 2008, *Regulation of the People's Republic of China on the Disclosure of Government Information* was officially put into implementation and ministries and their affiliated agencies and provincial governments announced the information publication procedures in accordance with applications, so the governmental information publication was more regulated and improved. www.gov.cn integrated the open governmental information directory systems and comprehensively published all kinds of documents formulated by the Central Government and the power list of the organizations of the Central Government (See Fig. 4.2). Basis setting, service guide and examination and approval procedures are provided for each administrative licensing, which plays a role of supervision for legal, transparent and regulated administration. By December 2016, 100% of the administratively licensed affairs in 31 provinces (autonomous regions and municipalities directly under the Central Government) were open; in 80.41% of them, the texts concerning application can be downloaded; application materials of 62.33% can be submitted online for two-way pre-examination and 56.47% can be finished online on the basis of interactive service.[11] Through the Internet the Chinese Government published information concerning bulletins, conferences, personnel appointment and removal, financial budget and settlement, and corruption combat, reports on the conferences or activities of leaders, governmental or departmental conferences, press release, and great events of the government or departments. To sum up, great progress has been made in information publication.

2. **Information sharing between governmental departments is being promoted.**

China tries to break the data barricade between governmental departments by setting up the unified data openness platform to meet the public demand for obtaining, processing and using information. In September 2016, the State Council issued *Interim Measures for the Administration of Sharing of Government Information Resources*, requiring to accelerate the interconnection between governmental information systems and public data sharing to give full play to the role of governmental information resource sharing in deepening the reform, transforming the function and innovating administration. The *Report on the Work of the Government of 2017* points out that we should accelerate the interconnection between the information systems of the State Council and local governments to form a unified governmental service platform. In May 2017, the State Council issued the *Implementation Plan for Governmental Information System Integration*

[11]Source: *Report on China's E-government Development* (2017).

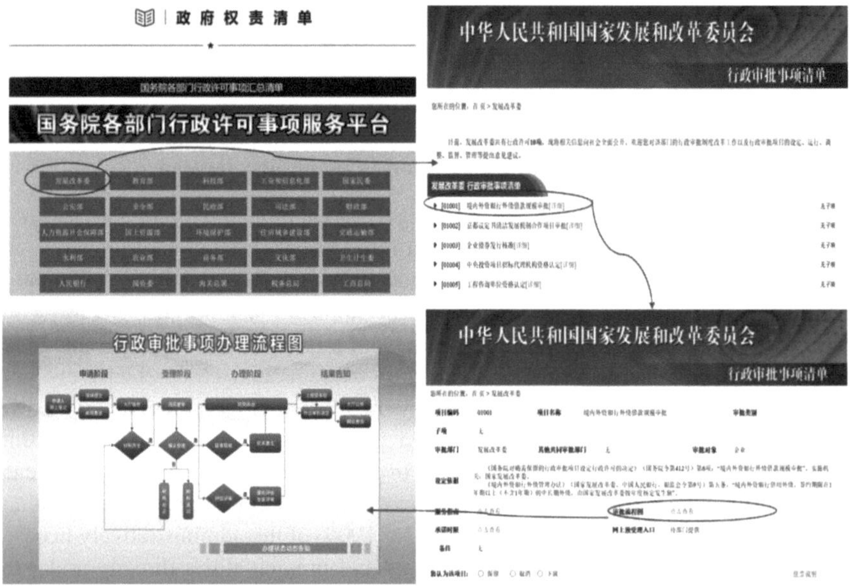

Fig. 4.2 List of power of the departments released on www.gov.cn

and Sharing, proposing the goal that breakthroughs should be made in key areas of governmental data sharing and openness. All regions and departments are active in putting the requirement into action to accelerate information sharing, and achievements have been made in this respect. By the end of 2016, there were 1.35 billion pieces of basic population information in the national population base, with the population photo acquisition saturation of 96.11% and the actual residential address saturation of 78.53%.[12] The information in the population base is shared by the departments of public security, education, civil affairs and human resources and social welfare, playing a great role in public security, social security and macro-control. The car license-plate lottery system in Beijing shares the information from the departments of public security, social security, taxation and civil affairs, avoiding repeated collection of information, reducing information collection cost and improving the information accuracy, to ensure the fairness, justice and transparency of the system.

3. **Streamlined administration and delegated authority are ensured.**

The one-stop e-government platform is a way of practicing streamlined administration and delegated authority. Governments of all levels have deepened the data resource opening and application, and strengthened data integration and resource value-added service, so that the public can better enjoy governmental service. The pilot reform of streamlined administration, delegated authority,

[12]Source: *Yearbook of China E-government* (2016).

combination of the two with strict management, and optimized service in Zhejiang Province aims to build a unified database of population, legal persons and e-licenses, so that the information can be shared by authorities of industrial and commercial administration, taxation, human resources and social security. By April 2016, there were 58.96 million pieces of shared population information, 2.49 million items of legal person data, and 8.6 million e-licenses of 37 departments. Besides, 560,000 business licenses of "five-in-one" had been issued through information sharing, and Zhejiang Construction Bureau had referred to the social security information through the e-licenses during the 110,000 qualification examinations and approvals,[13] so that the applying businesses and individuals can avoid being examined by other departments at the same time.

4.4 Emerging Information Technologies Are Making Governmental Service More Efficient

1. **Intensified construction of e-government based on cloud computing is being popularized fast.**

Application of e-government is transferring from separate construction to intensified construction based on cloud computing. According to statistics, by the end of 2016, in 15 provincial regions, the intensified construction of machine rooms, storage equipment and OA systems had succeeded based on "cloud". In Shaanxi Province, 1,233 application systems and hundreds of databases of 95 provincial departments, 11 prefectures or municipalities at the same level and 107 counties have been launched onto the e-government platform. Through intensified construction, 55% of fund and 60% of maintenance cost have been saved.

2. **Big data technology has been used to made decision making more scientific.**

Big data can directly reveal the connection of things while traditional technologies cannot. They make the macro-control decision-making more scientific and micro-management more precise. China Securities Regulatory Commission has combated "rat trading" and related transaction in securities transaction by using big data, which have helped financial and securities supervision authorities to improve their capabilities of tackling and supervising market risks and illegal transactions and thus enhance the surveillance. The "*jinshui sanqi* (Golden Taxation Third Stage)" Project launched by the State Administration of Taxation has centralized all the relevant application systems throughout the country. In that project, unified data standards and one voice have been adopted and key points of cross-provincial and cross-departmental circulation of tax-related data have been linked. Through the

[13]Source: Chen Lisan, *Internet + E-governmental Service: Practice of Zhejiang*.

deepening of big data sharing and utilization, tax source supervision has been improved, taxation procedures have been regulated and the criterion of law enforcement of taxation authorities in different places has been unified. In Shanghai, Shanxi and Chengdu, a comprehensive transportation information platform has been set up to integrate the transportation data of different sectors and organizations, and monitor the extreme traffic jam in real time to provide basis for the policemen to tackle the jam in time and for people to choose their route in a reasonable way and to provide big data support for urban planning.

3. **AI technology is used to improve intelligent governmental service.**

AI technology's application in e-government will transfer people-to-people service into machine-to-people service, and face-to-face service to non-meeting service. It will change the traditional service mode. By learning autonomously and applying contextual information of people, space and things, it can provide automatic identification and intelligent prediction, so it has been widely used in terrorism prevention and countering, environment monitoring and climate forecast. With the sharp increase of the demand for intelligent and personalized service, AI will be more widely used in e-government.

Chapter 5
Clean Cyberspace

Chinese Academy of Cyberspace Studies

General Secretary Xi has pointed out that the cyberspace is the common home of humankind. Making it better, cleaner and safer is the common responsibility of the international community, according with the interests of the people. Responsible for society and its people, China always strengthens the cyber content construction, spreads positive energy in cyberspace and cultivates healthy Internet culture. The country promotes integrated development of media, accelerates network communication law enforcement, and enhances cyberspace governance in accordance with laws, so that the cyberspace is full of more positive energy, rich culture and regulated order. All these have made the cyberspace cleaner and favorable to Internet users, especially juvenile users.

5.1 Healthy Public Opinion Ecosystem in Cyberspace

5.1.1 Network Media Have Become an Important Information Spread Channel

1. **There are more and more online news and information spreaders.**

Online news and information service has undergone three stages of development: web portals, social media and mobile media. There are an increasing number of news and information spreaders. On the very first day of 1997, people.cn was put into operation by *People's Daily*, the first central news website of China. Then Xinhua News Agency and CCTV also launched their websites providing news and information. A number of local news websites came into being and developed rapidly. Commercial websites like NetEase, Sohu and Sina were set up and their

Chinese Academy of Cyberspace Studies
Beijing, China

Chinese Academy of Cyberspace Studies (ed.), *China Internet Development Report 2017*,
https://doi.org/10.1007/978-3-662-57521-5_5

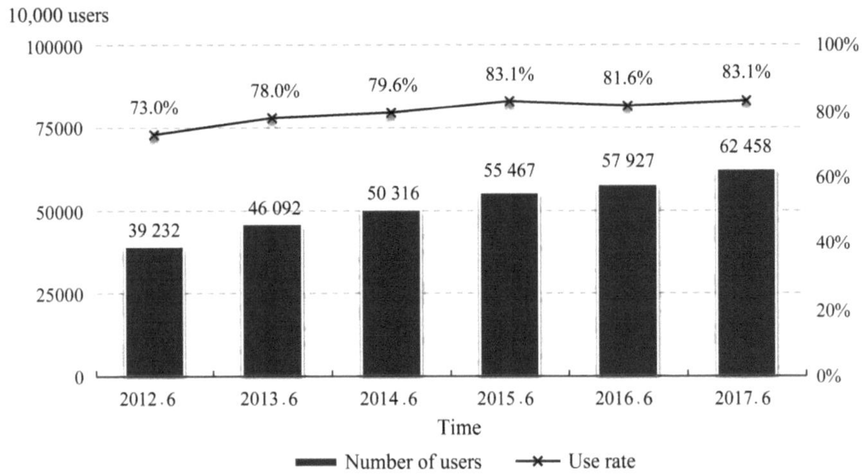

Fig. 5.1 Number of users and use rate of online news (*Source* CNNIC, Statistical Report on China Internet Development, June 2017)

news channels were launched. With the explosive growth of the mobile Internet market, news APPs represented by TouTiao.com and thepaper.cn and social media represented by Sina Weibo and WeChat have witnessed dramatic development. They are increasingly influential and have become important news and information channels. So far, in China, a news network layout has been formed with key news websites as the backbone and key commercial websites and news APPs as the active participants. By June 2017, there were 297 institutions with licensed qualification of Internet news and information service and 625 million online news readers[1] (See Fig. 5.1).

2. **The spread capacity of network media keeps improving.**

The rapid growth of the number of Internet news users and the dramatic improvement of Internet technology have enhanced the spread capacity of network media. Centering on the general situation of the CPC and the government, network media keep spreading positive energy and have launched a series of thematic promotion like "One Thousand Days after the Founding of the Central Leading Group of Overall Reform", promoting the situation, achievements and models. The Internet has become the main channel for introducing and studying the series of talks by General Secretary Xi Jinping and the new ideas, thoughts and strategies of the CPC Central Committee on the governance of China. During the 19th CPC National Congress, 15 central news websites formed the "national team" reporting the great event, playing a leading role in the news broadcast. According to statistics, the news reports on the 19th CPC National Congress were read 17 billion times and

[1]Source: CNNIC, *Statistical Report on China Internet Development*, June 2017.

the relevant videos were played 3.2 billion times.[2] Besides, during that period, the news spread approaches were innovated. The WeChat Official Account "News Center of the 19th CPC National Congress" was opened, the first WeChat official account in the Congress history, to meet the demand of Chinese and foreign journalists for interviewing and the demand of the public for knowing about the CPC's news and information through APPs. Meanwhile, network media make use of the advantages of online spread and adhere to online spread rules, playing a key role in reporting on hotspots of society and people's livelihood, big natural disasters and public emergencies. They release authoritative information and interpret policies and measures by losing no time. They also respond to social concerns, clear up confusion, remove doubts, clarify rumors and dredge public sentiments.

Column 15 Report on the V-day Parade at the Commemoration of the 70th Anniversary of the Victory of Chinese People's War of Resistance against Japanese Aggression and the World Anti-Fascist War

On September 3rd, 2015, the V-day parade was held in Beijing at the Commemoration of the 70th Anniversary of the Victory of Chinese People's War of Resistance against Japanese Aggression and the World Anti-Fascist War. Around one billion people around the world browsed and watched the commemoration and the parade online. There were over 20 million online reports on the event, with over 800 million page views and 100 million comments. The topic about the parade on Sina Weibo was read 5.1 billion times, with 8.56 million discussions; on WeChat, there were over 1.8 billion discussions and over 300 million page views of the parade video on it.[3]

3. **Network media spread the Chinese voice.**

Network media are the windows and channels for Chinese stories telling and Chinese voice spreading. They are active in innovating the external discourse system, following the cross-border Internet transmission trend and looking for cross-cultural transmission techniques. By enhancing the agenda setting and employing editors from overseas, they try to create new concepts, categories and expressions in a way acceptable and understandable to the overseas audience and improve the pertinence and effect of the news and information and to make them more inspiring. Meanwhile, people.cn, xinhuanet.com, cctv.com, china.com.cn, chinadaily.com.cn and cri.cn all broadcast news and audio & video programs to the world in different languages and have launched multi-lingual APPs. *People's Daily*,

[2]Source: Sina Weibo.

[3]Source: youth.cn, *One Billion Internet Users across the Globe Browse the V-day Parade and the Cyberspace is Surged with Patriotic Sentiments*, Beijing, September 4th, 2015.

Xianhua News Agency, CCTV and *China Daily* have opened their accounts on social network platforms overseas like Facebook, Twitter, YouTube, LINE and VKontakte (in Russia), attracting over 110 million followers. Their online international broadcast covers over 200 countries and regions, having one billion overseas users, with a growing acceptance rate.

5.1.2 Network Media Evolution and Innovation Have Boosted the Industrial Prosperity

1. **Integrated development of media has seen great progress.**

China keeps promoting the integration between traditional and new media in content, channels, platforms, operation and management and has thus formed a number of new transmission platforms with wide coverage and variety, advanced means and profound influence. In 2014, the Central Leading Group of Overall Reform passed the *Guiding Opinions on Promoting the Integration of Traditional Media and Emerging Media*, which proposes the directions, goals and approaches of the integrated development of traditional and new media. In July 2016, State Administration of Press, Publication, Radio, Film and Television issued the *Opinions on Further Accelerating the Integrated Development of Radio and Television Media and Emerging Media*, in which the goal is defined as "to form a wide variety of new mainstream media adopting advanced approaches with great competitiveness and to launch strong new media groups". Since then, mainstream traditional media like *People's Daily*, Xinhua News Agency and CCTV have enhanced their attention to new media and cultivated the model of "central kitchen". They have become the main force in facilitating the integrated development of media. By the beginning of 2017, the number of autonomous downloads of *People's Daily* APP had amounted to 154 million and the newspaper's legal person Weibo had had over 89 million followers on the platforms of people.cn, sina.com and Tencent, and its WeChat official account had had 8.4 million subscriptions; the number of downloads of the APP of Xinhua News Agency had reached 180 million, the number of its Weibo followers, 44.67 million and the number of its WeChat subscribers, 0.61 million; the number of the downloads of the APP Cbox of CCTV had reached 550 million, that of the downloads of its CCTV News APP, 41.95 million, that of its Weibo followers, 47million,and that of its WeChat subscriptions, 6.8 million.

2. **Mobile new media are producing increasing influence.**

In recent years, mobile newspapers, mobile WAP websites, mobile APPs, and official Weibo and WeChat official accounts have emerged one after another, producing increasing influence. By June 2017, the number of mobile phone news APP users in China had reached 596 million, accounting for 95.4% of the total number of network news readers (See Fig. 5.2). The use rate of WeChat moments

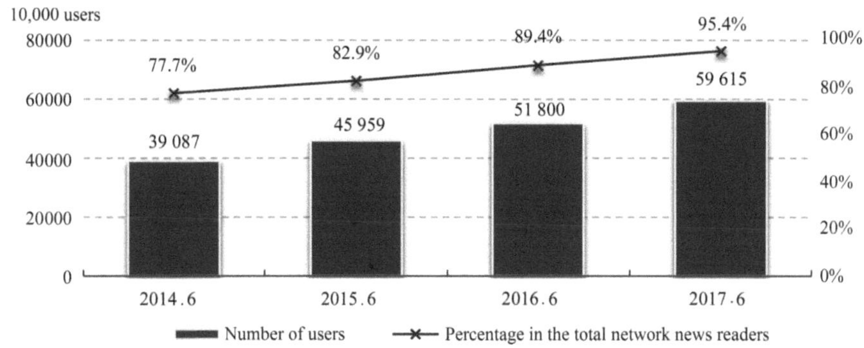

Fig. 5.2 Number of mobile phone newspaper users and the use rate in China (*Source* CNNIC, Statistical Report on China Internet Development, June 2017)

and Weibo had respectively reached 84.3% and 38.7%[4] by June 2017. The mobile terminal has become the leading source of news and information for Internet users, surpassing the PC terminal. Traditional media have launched their mobile phone media through mobile platforms, strengthening the integration between traditional and new media. By July 2017, the total number of mobile phone newspaper users throughout the country had reached 300, and that of "One Newspaper of Each Province" APP users, 200 million. A number of mobile newspaper media had had over 10 million users by that time. For instance, the number of Sichuan Mobile Newspaper users had increased from 13 million in 2013 to 68 million in 2016.[5]

5.1.3 Internet Socializing Is Increasingly Popular Among Internet Users

1. **Instant messaging has ushered in the new socializing mode in the Internet era.**

The rising of instant messaging and socializing service on the Internet has totally changed the communication method and thinking of the Chinese. "No socializing, no life" has become the new normal of Internet users' lifestyle in the country. The earliest instant messaging tool in China was OICQ developed by Tencent and launched online in March 1999. From then on, instant messaging applications have emerged one after another, with more and more functions. With the fast

---

[4]*Deep Integration to Form a New Layout of Media*, http://politics.people.com.cn/n1/2017/0105/c1001-28999350.html.

[5]Office of the Central Leading Group for Cyberspace Affairs: To Promote the Innovative Development of New Media like Mobile Phone Newspaper throughout the Country, http://news.sohu.com/20160713/n458959821.shtml.

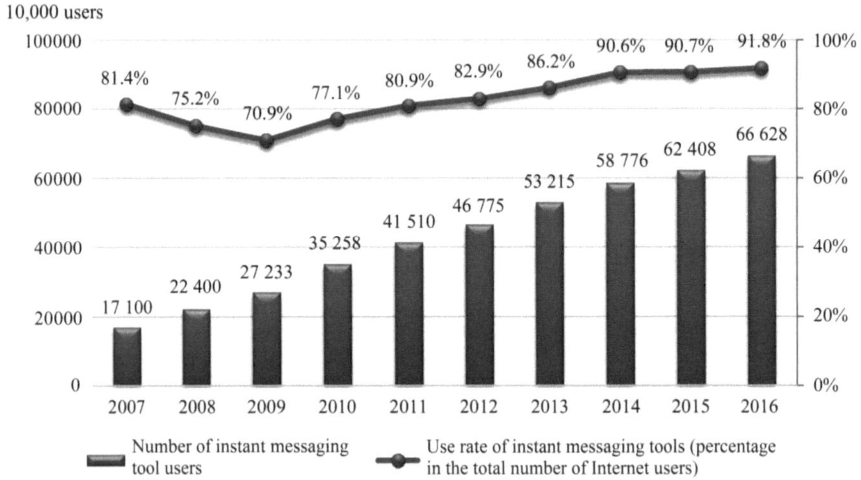

Fig. 5.3 Number of instant message tool users and the use rate in China (2007–2016) (*Source* CNNIC, Statistical Report on China Internet Development, June 2017)

Table 5.1 Representatives of different areas of mobile social apps in China

Applications		Representative APPs
Instant messaging tools		QQ, WeChat, Fetion and easychat
Comprehensive social applications		Sina Weibo, Tencent Weibo and Qzone
Vertical social applications	Community	Post bar, douban and zhihu
	Entertainment	Meitu, Meipai, YY and Mstching
	Anonymous	Wumii, Llspace and Momo
	Match-making	jiayuan.com and youyuan.com
	Career	LinkedIn and maimai

development of mobile applications like WeChat and the launch of mobile phone versions of mobile social websites, online socializing has shifted from the PC to mobile. Today, China has a complete market system of social network applications, covering instant messaging tools like QQ and WeChat, comprehensive social APPs like Sina Weibo and vertical social APPs with segmented fields. This market is expanding into industrial chains of pan-entertainment, social marketing, Internet finance and big data. By the end of June 2017, the number of instant messaging tool users in China had reached 692 million.[6] The number of instant messaging tool users and their use rate in China from 2007 to 2016 is shown in Fig. 5.3, and the representatives of different areas of mobile social APPs in China are shown in Table 5.1.

[6]Source: CNNIC *Statistical Report on China Internet Development*, June 2017.

2. **Social network is developing into an eco-platform "connecting everything".**

Social APPs are emerging one after another with more and more functions. Instant communication, news broadcast, live streaming, payment transaction, game playing and public service can all be done on social APPs, with the platform development covering multiple fields and enhancing cohesion among users. Take WeChat as an example. Besides instant messaging, WeChat official account, wallet and applet have all been launched to form a complete ecosystem. By 2016, the number of monthly active accounts of WeChat Pay and daily payments through WeChat Pay had respectively amounted to over 600 million.[7] Social networks have also sped up the diversification of Internet business mode. Marketing and mobile advertising based on social networks have become the most active areas. Combined with social circles and location service, online marketing is more precise and personalized, so it is a new growth point in e-commerce.

Column 16 WeChat: Connecting Everything

On January 21st, 2011, Weixin (WeChat) was launched by Tencent as a mobile instant messaging tool, a typical one in the era of the mobile Internet. It took less than one year for Weixin to have more than one million users after the whole system was completely opened. It supported cross-telecommunication-operator, cross-operation-system and cross-platform information transmission, as well as voice, message video, audio, picture and text transmission. In April 2012, Tencent launched the overseas version of Weixin 4.0, named WeChat, supporting multilingual communication. WeChat then became a successful model in Chinese models' going overseas. By the third quarter of 2017, there had been 980 million monthly active users of WeChat, over 100 million of whom were overseas. Among the monthly active users, one third spend over four hours on WeChat every day; and the average time spent on WeChat APP per capita is 63 min. WeChat has developed faster than QQ, ever faster than MiTalk, momo and easychat, into China's leading mobile social tool. After years of development, instant messaging tools represented by WeChat today are not only communication tools, but also providers of online and offline services. They are mobile platforms combining socializing, information, business, service and entertainment. They have formed complete ecosystems with IM as the entrance, covering all areas of the users' life and creating a new mode of online social life.

[7]Source: *Annual Performance Report of Tencent* (2016).

3. **Social media are producing increasing influence on public opinions.**

Social media have become one of the most popular types of Internet media. Characterized by a large user base, fast information transmission and strong function of interaction, they are a key force influencing public opinions on the Internet. Of course, traditional media are active in embracing social networks by opening their Weibo and WeChat official accounts to release important information, expand their transmission scope and enhance their public opinion influence. Besides, "We Media" are producing increasing influence. By June 2017, on the WeChat official platform, there had been over 28 million official accounts, among which there were 514,000 governmental accounts and 251,000 news accounts.[8] It is common for relevant contents to be read for over 100,000 times. By the second quarter of 2017, the number of monthly active users of Sina Weibo had amounted to 361 million.[9] The hot topics that are changed every 10 min produce great influence on public opinions on the Internet. Identity-verified Weibo users and online celebrities have greater influence on online topics. While making communication convenient, social networks have given rise to problems, such as harmful information spread, privacy disclosure and excessive reliance on the networks. Moreover, they have increasing capacity of social mobilization, so they pose great challenges to cyber governance.

5.2 The Legal System Concerning Internet Is Being Improved

General Secretary Xi Jinping points out that cyberspace is not a place beyond the rule of law. It is virtual, but players in it are real. Everyone should abide by the law, with the rights and obligations of parties concerned clearly defined. Since the 18th National Congress of the Communist Party of China, the Chinese Government has been speeding up the legislation on cyberspace. The country has launched a basic law in this filed, i.e. *Cybersecurity Law of the People's Republic of China*, and a series of other laws and regulations on Internet news and information service and released judicial interpretations concerning the cyberspace and made the present laws applicable to cyberspace, thus providing solid legal guarantee for the orderly and healthy development of the Internet.

[8]Source: CNNIC, *39*th *Statistical Report on China Internet Development.*

[9]Source: CNNIC, *39*th *Statistical Report on China Internet Development.*

5.2.1 The Legislation on Cyberspace Is Speeding up

1. ***Cybersecurity Law of the People's Republic of China* is the basic law of the cyberspace.**

On November 7th, 2016, the Standing Committee of the National People's Congress passed the *Cybersecurity Law of the People's Republic of China*, which came into force on June 1st, 2017. It is China's first basic law comprehensively regulating cybersecurity, so it is a milestone in the legislation on cyberspace. The law, for the first time, defines the security obligations of online product and service providers and network operators, perfects the rules of private information protection, and clarifies the overall basic institution concerning the national cybersecurity areas like protection of critical information infrastructure security and cross-border transmission of personal information and key data. It will become a legal weapon of governing the Internet in accordance with law and eliminating cyber risks, as well as a guarantee for the legal development of the Internet.

2. **Legislation on relevant areas of cybersecurity is progressing.**

In August 2004, China issued *Electronic Signature Law*, which defines the legal effect of electronic signature. From 2014 to 2015, *Law of the People's Republic of China on Protection of Consumer Rights and Interests*, *Food Safety Law* and *Advertising Law* modified by China are regulations on online sales, food purchase and advertising. *Amendment (IX) to the Criminal Law of the People's Republic of China* lists online rumor spread and privacy disclosure into the *Criminal Law*, improving the deterrence of cyber governance in accordance with laws. In November 2017, the Standing Committee of the National People's Congress passed the amended *Anti-Unfair Competition Law*, which is a special regulation on illegal competition like *shuadan* (false transaction) and credit speculation in e-commerce.

3. **The judicial interpretations of the Supreme People's Court and the Supreme People's Procuratorate make laws applicable to cyberspace governance.**

To conform to the cyberspace governance demand and characteristics, the Supreme People's Court and the Supreme People's Procuratorate have launched a series of judicial interpretations concerning the Internet to make laws applicable to cyberspace governance and thus to improve the Internet law system of China. Those judicial interpretations are as follows: *Interpretation of the Supreme People's Court on Application of Laws in the Trial of Civil Disputes Over Domain Names of Computer Network*, *Interpretation of the Supreme People's Court and the Supreme People's Procuratorate on Several Issues concerning the Concrete Application of Law in the Handling of Criminal Cases of Making, Reproducing, Publishing, Selling and Spreading Pornographic Electronic Information by Means of the Internet, Terminal of Mobile Communications and Sound Message Stations*, *Interpretation of the Supreme People's Court and the Supreme People's Procuratorate of Several Issues on the Application of Law in the Handling of*

Criminal Cases about Endangering the Security of Computer Information Systems, Interpretation of the Supreme People's Court on Several Issues concerning the Application of Law in Hearing Civil Dispute Cases Involving Infringement of the Right of Dissemination on Information Networks, Interpretation of the Supreme People's Court and the Supreme People's Procuratorate on Several Issues concerning the Specific Application of Law in the Handling of Defamation through Information Networks and Other Criminal Cases, Provisions of the Supreme People's Court on Several Issues concerning the Application of Law in the Trial of Cases involving Civil Disputes over Infringements upon Personal Rights, Interests through Information Networks and *Opinions of the Supreme People's Court and the Supreme People's Procuratorate on Several Issues concerning the Application of Law in the Handling of Telecommunications Network Fraud and other Criminal Cases*.

5.2.2 Internet Policies and Regulations Have Been Launched

At the early stage of Internet development of China, Internet access administration was the core of policy and regulation making, covering domain name management and international networking. On February 1st, 1996, the State Council released *Provisional Regulations of the People's Republic of China on the Management of International Networking of Computer Information Networks*, which was the first regulatory document in the nature of law after China's access to the Internet. On June 3rd, 1997, China Internet Network Information Center (CNNIC) was founded and later *Provisional Measures to Administer the Registration of Domain Names* and *Implementing Rules of Domain Name Registration* were released, marking that China began to strengthen domain name administration from the perspective of laws and regulations. In 1998, four ministries, namely, Ministry of Public Security, Ministry of Industry and Information Technology, Ministry of Culture and State Administration of Industry & Commerce, jointly launched *Notice on Regulating the Operation of Internet Bars and Strengthening Their Security Management*, in which the qualifications of Internet bar operators and examination and approval procedures for the bars are defined. At the early stage, cyber information security began to be channeled into policy topics and the State Council, Ministry of Public Security and Ministry of Personnel (today's Ministry of Human Resources and Social Security) released relevant regulations and rules, laying a foundation for cyber security and administration.

After 2000, regulating Internet application became the focus of policy and regulation making, covering such areas as news transmission, publishing, intellectual property right, games, audio/video show and medical care, e-commerce, e-payment, e-taxation and civil disputes among Internet users. In 2000, the State Council issued *Regulations on Telecommunications of the People's Republic of*

China and *Regulations on Internet Information Service of the People's Republic of China* to regulate the telecommunication market and Internet information service. On September 29th, 2002, the State Council issued *Regulations on the Administration of Business Sites of Internet Access Services*, which defines the qualifications and obligations of Internet bar operators and the forbidden actions and relevant punishments. From 2004 to 2006, a series of policies were formulated to form a basic regulatory system for Internet application, including *Self-discipline Rules of Spreading Pornographic and Other Harmful Information on Websites*, *Measures for the Administration of the Publication of Audio-Visual Programs through the Internet or Other Information Network*, *Measures Regarding the Administration of Drug Information Service over the Internet*, *Measures for the Archival Administration of Non-operational Internet Information Services*, *Measures for the Administration of IP Address Archiving*, *Measures for the Administration of Electronic Certification Services*, *Measures for the Administrative Protection of Internet Copyright*, *Regulations on the Protection of the Right to Communicate Works to the Public over Information Networks* and *Provisions for the Administration of Internet News Information Services*. On June 25th, 2007, *Plan for the Development of E-commerce during the 11th Five-year Plan Period* was launched, the very first e-commerce development plan of China. Then e-commerce developed from a typical Internet application into the key emerging industry of the country, taking the lead among Internet applications.

To promote healthy and orderly development of the Internet, China persists in simultaneous encouragement, support and regulation as well as guidance by policies and laws. A series of regulatory documents have been launched to regulate online shopping, social media, Internet finance, online advertising, live streaming and online car-hailing. In January 2017, the General Office of the CPC Central Committee and the General Office of the State Council issued the *Opinions on Promoting the Sound and Orderly Development of the Mobile Internet*, making layout for the orderly and healthy development of the mobile Internet. The Cyberspace Administration of China have launched a series of cyberspace administration regulations, including *Interim Provisions on the Administration of the Development of Public Information Services Provided through Instant Messaging Tools*, *Provisions on the Administration of Account Names of Internet Users*, *Provisions on the Interview of Entities Providing Internet News Information Services*, *Provisions for the Administration of Internet News Information Services*, *Provisions on the Administration of Internet Information Search Services*, *Provisions on the Administration of Mobile Internet Applications Information Services*, *Provisions on the Administration of Internet Live-Streaming Services*, *Provisions on the Administration of Internet Comments Posting Services*, *Provisions on the Administration of Internet Forum and Community Services*, *Rules to Regulate Internet Group Information Services*, *Provisions on the Administration of Internet User Public Account Information Services*, *Provisions on the Administration of the Safety Assessment of New Technologies* and *Applications for Internet News Information Services and Measures for the Administration of Content Management Practitioners in Internet News Information Service Providers.*

Ministry of Commerce, Ministry of Transport, Cyberspace Administration of China, State Administration of Industry & Commerce, "one bank and three commissions" (namely, the People's Bank of China, China Banking Regulatory Commission, China Securities Regulatory Commission and China Insurance Regulatory Commission) and some trade associations have also formulated a number of regulations as compensations for the laws to regulate the development of the Internet. These regulations are *Guiding Opinions on Promoting Healthy Growth of Online Shopping*, *Interim Measures for the Administration of Online Taxi Booking Operations and Services*, *Interim Measures for the Administration of Internet-Based Advertising* and *Guiding Opinions on Promoting the Sound Development of Internet Finance*.

5.2.3 Cyberspace Law Enforcement Has Been Regulated and Regularized

1. **The law enforcement system of cyber content administration has been established and improved.**

Since 2013, construction of the three-level system of the cyber law enforcement has been progressed throughout the country and the mechanism has been improved. The coordination mechanism for administrative law enforcement has been set up among 12 departments, including industry and information technology, public security, and industry and commerce and so on. The interview, urge and warning mechanism is being set up. Cyber law enforcement regulatory construction is promoted across the country, special training is carried out in that respect and new cyber content administration licenses are issued. In May 2017, Cyberspace Administration of China launched *Provisions on the Administrative Law Enforcement Procedures for Internet Information Content Management* to unify the evidence standard for law enforcement and thus regulate law enforcement. In September 2017, Beijing and Guangzhou cyberspace administration authorities execute administrative punishment to Sina Weibo, Baidu and Tencent since there had occurred violence and terrorism information, false information and pornographic information on these platforms.

2. **Cybercrime keeps being combated.**

To counter cyber fraud and privacy violation, China keeps enhancing law enforcement and has organized a number of actions, which have been successful. Typical cases have been investigated and handled in accordance with laws. For instance, Joint Investigation Unit for Wei Zexi Incident required Baidu to rectify its ranking mechanism based on payment for search; in Xu Yuyu Case, the entire cyber fraud gang were arrested; the fund raised in the "Pseudo Charity" and "Luo Yixiao" cases at kuaishou.com has been returned. Departments of cyberspace administration, industry and information technology, public security, and industry and commerce have launched special harness actions targeting search engine and online advertising to crack down privacy disclosure and telecommunication fraud. In March 2017, Ministry of Public Security began countering hackers' attack and privacy violation. By July that year, 1,800 above-mentioned cases had been detected and over 4,800 criminals had been arrested, involving over 50 billion pieces of private information.[10]

3. **Harmful information is being eliminated.**

As for cyber rumors, terrorism and harmful information concerning ethnic groups and religions, authorities of cybersecurity, public security, and industry and information technology cooperate with each other in countering online extortion and paid post deletion, combating crimes from pseudo base-station equipment, and rectifying regulation violation and credit loss of match-making websites. They have also jointly launched a series of actions like Cleaning, Seedling-protection and Sword Net to clean up the cyberspace. In June 2015, Ministry of Public Security called on all the public security authorities throughout the country to establish the mechanism for regular public inspection by net police. By the end of 2016, 371 net police inspection agencies had found out and tackled over 1.1 million pieces of harmful information.[11] In August 2017, eight departments including National Office against Pornography and Illegal Publications started their 40-day specific rectification of vulgar and obscene information in the cyberspace. By September that year, over 20 million pieces of such information had been found out and disposed of.[12] In the year 2016, cyberspace administration authorities throughout the country interviewed 678 illegal websites in accordance with laws. In the same year, 3,467 illegal websites were deprived of their licenses or records or even were closed by cyberspace administration authorities together with industry and information technology authorities.[13]

[10]Source: *Progress Has Been Made in the Specific Action of Combating Hackers' Attack and Privacy Violation*, http://www.mps.gov.cn/n2253534/n2253535/n2253537/c5741823/content.html , March 10th, 2017.

[11]Source: *Two years of Public Security Reform: capacity of cyber governance keeps increasing*, http://www.mps.gov.cn/n2255079/n4876594/n4974590/n4974593/c5635761/content.html, February 13th, 2017.

[12]Source: *Pornographic Information Has Been Completely Checked in Cyberspace and Staged Achievements have been Made in the Special Action,* http://www.shdf.gov.cn/shdf/contents/767/350038.html, September 25th, 2017.

[13]Source: *Strict Law Enforcement Forms Deterrence, and Regulation Consolidates the Foundation: sustaining progress in administrative law enforcement by cybersecurity administration in 2016*, http://www.cac.gov.cn/2017-01/20/c_1120352553.htm. January 20th, 2017.

5.2.4 Cyberspace Jurisdiction Is Being Innovated

China is facilitating its cyberspace administration in accordance with laws, seeking to innovate relevant ideas and measures through new technologies and applications of the Internet, so that administration innovation can keep up with innovative Internet development. In this respect, the country has seen a number of good practices and improved effect of management. E-court has been set up by the Supreme People's Court as a pilot project. By November 2016, 3,520 courts, and 9,277 people's tribunals and maritime courts had been connected into the Court Network.[14] Thus all courts throughout the country use one "net" to handle the cases and daily affairs, and "move" lawsuits from offline to online, solving the problem of "more cases but fewer officers". In March 2015, the Lawyer-serving Platform of Guangzhou Arbitration Commission, the very first of its kind in China, was launched in Guangzhou. It has more than 30 functions, including online registration, case query and statistics, and judgement document download and so on. It is a cross-regional platform which can handle arbitration of different places online. In August 2015, the online e-commerce court of Zhejiang People's Court was officially put into operation. By mid-June of 2017, it had received 22,787 case applications and its average court session lasts half an hour.[15] On August 18th, 2017, Hangzhou Internet Court was founded as the country's first pilot court handling online cases. It is a typical model reflecting the idea of "online tackling of online cases", a combination of Internet technology and mechanism innovation specially dealing with online disputes.

Column 17 China's First Internet Court Founded in Hangzhou

On June 26th, 2017, the Leading Group for Overall Reform deliberated and passed *Plan on Founding Hangzhou Court of the Internet*. On August 8th, the Supreme People's Court announced that Hangzhou Internet Court as a pilot specially handling Internet cases would be founded on the basis of Hangzhou Railway Transportation Court. On August 18th, 2017, the prosecutor in Hangzhou and the defendant in Beijing had their trial on the Internet. It was the case of Zhenhuan's Accusation of Netease Online. The session only lasted 20 min before both parties agreed on mediation. On the same day, Hangzhou Court of the Internet, the first of its kind in China, was founded. It is engaged in handling Internet cases under the jurisdiction of the people's courts within Hangzhou. In this court, every step is done on the Internet, including prosecution, case filing, service, testification, session opening and judgement. All

[14]Source: *Blue Paper of Rule of Law: Report on Informatization of Chinese Courts No.1 (2017).*

[15]Source: Wang Chun, journalist from legaldaily.com.cn, *Introduction to Hangzhou Internet Court*, Legal Daily—legaldaily.com.cn (June 29, 2017).

the action steps are recorded immediately and continuously and the parties can see their case handled without spending any time or money on the trip.[16]

The founding of Hangzhou Court of the Internet marked the great reform in rule of law over the cyberspace and the real integration between Internet governance and its utilization. The court uses big data analysis technology to make multi-module contrastive analysis of data concerning the case, and summarizes the rules and characteristics to form structured and standardized Internet judgement rules. Thus the multi-level diversified dispute settlement system is formed involving precedent resolution under guidance, online dispute resolution (ODR) and third-party mediation and litigation, to create a more secure, fair, efficient and personalized judicial system concerning cyberspace.

5.3 The Diversified Cyberspace Governance Ecosystem Is Being Constructed

Through laws and regulations, administration, industrial policies, technical standards, promotion and education, the Chinese Government coordinates all parties' active participation in Internet governance, including Internet businesses, social organizations, specialists, think tanks and Internet users, who are a great force in promoting the development of the diversified Internet governance ecosystem. Thus an Internet governance model conforming to the rules of the Internet itself with the Chinese characteristics is being established.

5.3.1 Businesses Play Principal Roles in Internet Governance

1. **Businesses' responsibilities are strictly fulfilled.**

Chinese Internet businesses make active response to the Government's measures on Internet governance, trying to fulfill their responsibilities. In 2013, search engine businesses such as Baidu and Qihoo 360 issued their plan for guaranteeing their users' rights and interests, promising to make compensations for the loss that may be caused to their users by malicious fraud information and phishing sites. In accordance with the requirements for their main responsibilities, Internet businesses have made and improved their management and tackle any illegal or rule-breaking

---

[16]Xu Jun, journalist from People's Daily, China's First Internet Court is Founded in Hangzhou and Lawsuits can be Handled Keyboard to Keyboard, Column of Rule of Law, People's Daily (September 6, 2017).

accounts and harmful information on Weibo, Blog, WeChat, BBS and Tieba. To fulfill the requirements of National Copyright Administration,cloud storage service providers like Baidu Netdisk, Sina vdisk and UC netdisk did overall cleaning of the harmful information concerning piracy and patent infringement in 2016. Telecommunication operators like China Telecom, China Mobile and China Unicom have enhanced their governance technology improvement and system and mechanism building, making their effort to cut off false calls. They have launched reminders of preventing false messages and countered pseudo base-station defraud. Meanwhile, Internet businesses maintain the order on cyberspace in accordance with laws and regulations. Online Game Anti-Indulged System developed by Tencent and the Identity Verification System developed through cooperative Android game APPs show that at the stage of design of all products, the Internet businesses already consider the key points and difficulties in governance and set up their self-discipline mechanism.

2. **Self-discipline organizations have been founded.**

Internet businesses have founded self-discipline alliances in different fields at different levels to discipline themselves. In November 2014, Sina, Sohu and NetEase published the self-disciplinary rules of the Honesty and Self-discipline Alliance of the Internet Industry. Chinese Internet businesses initiatively take the principal responsibilities for maintaining the market order. In November 2012, 12 Internet businesses including Baidu, Jike, Panguweb, Qihoo 360, Shanda Literature and Sogou signed *Self-discipline Convention of Internet Search Engine Service*, jointly proposing to resist the transmission of illegal and harmful information, pursue the trade conventions and business rules, create favorable competition environment, and protect intellectual property right, privacy and personal information security. In June 2013, 21 Internet businesses including Alibaba, Tencent, Baidu, Sina, Shanda and NetEase founded Internet Anti-Fraud Committee to counter fraud jointly and form the strategic cooperation framework of joint prevention of and combat against illegal actions against the e-business ecosystem.

5.3.2 Internet Societies Are Playing Their Roles

In May 2001, Internet Society of China released *Self-discipline Convention on the Internet Industry of China*, *Specification of Internet Society of China against Spam* and *Self-discipline Convention on the Network Copyright on the Internet of China* and set up Anti-Spam Coordination Group, Working Committee of Industrial Self-discipline and Cyberspace Copyright Alliance to promote the healthy development of the Internet. In August 2011, China Netcasting Services Association was founded, who released *Self-discipline Convention of China Netcasting Services Association* and *Proposal on Netcasting Services Stressing Civility and New Practice* to resist harmful information spread, combat piracy and advocate copyright protection. In December 2012, Internet Access Service Association of China was founded to regulate the order in Internet bars, ensure the legal rights and interests of the Internet bar runners and Internet users and promote the transformation and upgrading of Internet service. In April 2015, Chinese Culture Institute of Internet Communication was founded to promote the innovative development of excellent Chinese culture in the era of Internet and speed up the introduction of excellent Chinese culture to the world. In August 2015, China Internet Development Foundation, the first public fund of Internet in the world, was founded in Beijing to integrate social resources, mobilize social forces and provide financial aid for the healthy development of the Internet. In March 2016, Cybersecurity Association of China and National Internet Finance Association of China were founded simultaneously, who respectively play an active role in mobilizing social forces to participate in maintaining cybersecurity and guiding the development of Internet finance. In recent years, a number of local Internet organizations like Beijing Internet Association have been founded and they have launched a number of activities advocating network civilization and thus promoted the formation of network literacy. On June 7th, 2017, the first promotion conference of Internet organizations of the whole country laid out the "Concentric Circles" Project to be implemented by the Internet organizations and announced 100 brand projects (2016–2017, the first group) of the Internet organizations nationwide. By December 2016, there were 1,333 Internet organizations throughout the country, with an increase of 144% in comparison with the number in 2015. Among all these organizations there are 47 national ones and 1,286 local ones,[17] with Zhejiang, Jiangsu, Sichuan and Guangdong having the leading numbers, exceeding 100 respectively.

5.3.3 Internet Users' Network Literacy Is Being Improved

1. **The public are more active in participating in Internet governance.**

China mobilizes the people's initiative to participate in Internet governance to clean up the cyberspace and maintain its order. The public supervision is an indispensable part in the diversified ecosystem of Internet governance. Since 2004, China has set up Internet Crime and Harmful Information Reporting Center, Internet Crime Reporting Center, Harmful and Junk Online Information Reporting and Acceptance Center (12321) and Pornography, Illegal Publications and Copyright Infringement Reporting Center (12390) to improve the working mechanism for illegal and rule-breaking information report and encourage the public to participate in Internet supervision and governance. In 2016, there were 40 million reports on Internet and 92% of harmful information was disposed of.[18]

---

[17]Source: Data released at the first promotion conference of Internet organizations (June 7th, 2017).

[18]Source: China Internet Crime and Harmful Information Reporting Center.

In 2015, China launched the series of activity called "Good Internet Users in China" and in early 2016, called on the Internet users to discipline themselves. Then education was carried out on network literacy through classes on Internet and public-interest advertisement contests and other theme events. In the two sessions of National Cyber Security Publicity Weeks, forums on network literacy education were held, so that the idea of abiding by laws, expressing oneself rationally and browsing the Internet in a refined manner can be embraced and practiced by Internet users.

2. **There have emerged civil cyberspace supervision organizations.**

While resisting illegal online behaviors and participating in online civilization activities, Chinese Internet users have set up their cyberspace supervision organizations of different types, actively involving themselves in building civilized, green and harmonious cyberspace. Cyberspace supervision organizations such as Mom Jury, Cyberspace Mom Monitoring Group and Rumor Refuting Alliance have made their effort in protecting juveniles' browsing the Internet, setting up the supervision mechanism for Internet users' feedback and rectifying online transmission. They have contributed to cyberspace clean-up and public opinion ecosystem construction.

Column 18 Chinese Internet Users Spontaneously Found Cyberspace Supervision Organizations

Zong Chunshan, the initiator of Mom Jury, is Director of Beijing Youth Legal and Psychological Consultation Service Center. Mom Jury is made up mothers recruited, who, pursuing the principle of the juveniles' interest being the priority, report and review the cyber content affecting the physical and mental health of juveniles, and submit their disposal suggestions to competent authorities and supervise the final handling of the cases. The Jury acts as a bridge for the public to participate in the supervision over and review of unfavorable online information.

Rumor Refuting Alliance, initiated by Dou Hanzhang on May 18th, 2011, is China's first civil organization of online rumor refuting. In this alliance (with their Sina Weibo account "辟谣联盟 (Rumor Refuting Alliance)" are several thousand enthusiastic Internet users who are active in clarifying facts and unearthing the truth). In the complicated environment with "we media", Rumor Refuting Alliance is not only a self-disciplinary organization, but also undertakes the responsibility of cleaning up the Weibo environment. Therefore, it is an innovative model of Internet users' participation in Internet governance.

Chapter 6
Improving Capacity of Cyber Security Safeguarding

Chinese Academy of Cyberspace Studies

Without cyber security there would be no national security and without informatization there would be no modernization. Cyberspace has become the fifth space, coming after the land, the sea, the outer space and the sky. It is the focus of leading countries' competition and the strategic area for maintaining the security of a nation. Guided by the principle of "establishing a correct philosophy on cyber security" proposed by General Secretary Xi Jinping, China aims to ensure development through security and promote security through development, persisting in the equality between cyber security administration and technical support. It tries to establish the complete cyber security review mechanism, enhance the security protection of critical information infrastructure and keep improving the people's awareness and skills of cyber security protection to build a solid cyber security defense.

6.1 General Situation of Cyber Security Is Serious

The new-generation information technology is penetrating all areas fast and the cyber security environment is becoming increasingly complicated. There are more cyber security risks and frequent incidents concerning critical information infrastructure, personal information and the mobile Internet. There are an increasing number of cyber attacks like APT, DDoS, Trojan and botnet, and ransomware. Therefore, the general situation of cyber security in China is serious.

Chinese Academy of Cyberspace Studies
Beijing, China

Chinese Academy of Cyberspace Studies (ed.), *China Internet Development Report 2017*,
https://doi.org/10.1007/978-3-662-57521-5_6

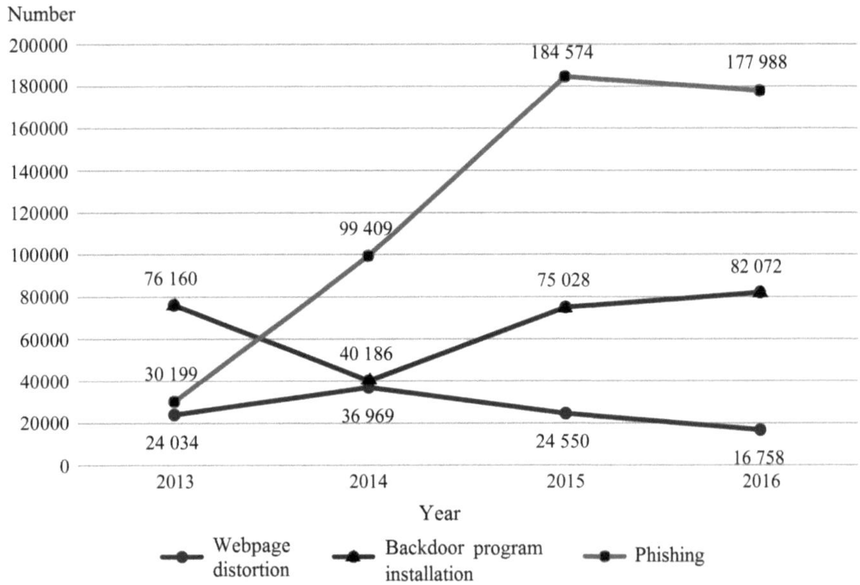

Fig. 6.1 Website security monitoring (2013–2016) (*Source* CNCERT)

6.1.1 Cyber Security Risks Are Prominent in All Areas

1. **Website security problems occur frequently.**

In recent years, there have frequently occurred web page distortion, backdoor program installation and phishing. Especially after the update service of Windows XP was cancelled in 2014, a large number of Windows XP users were faced with realistic and potential security risks. According to statistics from CNCERT, in the first three quarters of 2017, there were 17,353 websites that had been distorted in China, including 543 governmental ones; 24,353 websites were installed with backdoor programs, including 1,191 governmental ones; there were 16,099 phishing webpages, involving 4,026 IP addresses, fewer than those in 2016. Besides, vulnerabilities in basic applications and hard and software give rise to serious threats, for example, the remote code execution vulnerability Apache Struts2 S2-045 found in March 2017. The attacker could usurp the control of the server through the vulnerability and then 60.1% of the websites adopting the framework were affected. Website security monitoring from 2013 to 2016 is shown in Fig. 6.1.

Column 19 Means of Cyberattack
Website distortion refers to malicious damage or distortion of the content on the webpage so that the website stops working or there occurs abnormal content that hackers have put in. Website distortion can be categorized into dominant and recessive distortions. Through the former, the hacker can show

off his/her technique or announce his/her purpose. The latter means that the attacked website is installed with an invisible link with illegal information like pornography and fraud so that the hacker can get some financial benefits that are in fact illegal. To distort a website, the hacker has to know about the vulnerability of the website and install a backdoor program and finally usurp the control over the website.

Website backdoor installation refers to the hacker's uploading a remote-control page into certain directory of a website to control its server. The hacker can upload, browse, alter and delete the documents on the website, read and alter the data in the database or even run the system command on the server.

Phishing is the typical application of the combination between social engineering deception and network technology. In this attack, a website similar with the targeted one is constructed to trick the users. Phishing websites usually transmit information through spam, live chat, mobile phone messages or false advertising. Users may have their account number, password or other private information disclosed when they visit a phishing website.

2. **There are an increasing number of security risks for critical information infrastructure.**

With the rapid development of "Internet+" and intelligent manufacturing, the Internet is penetrating all fields of social life fast. Critical information infrastructure– information systems or industrial control systems in key industry sectors in particular– is becoming the key area subject to cyberattack. Any security incident will cause serious damage to the country and its people. By the end of September 2017, China National Vulnerability Database (CNVD) had recorded 1,313 industrial control vulnerabilities, including 307 in 2017, the number having increased by 77.5% in comparison with that in 2016. Major industrial control vulnerability risks include buffer overflow, lack of access control mechanisms, weak password and directory traversal.[1] CNCERT has found 72,767 pieces industrial control equipment connected into the Internet, the protocols involving S7Comm, Modbus, SNMP, EtherNetIP, Fox, FINS and BACnet, and manufacturers including Siemens, Rockwell, Schneider, Omron, Wind River and Vykon Controls and hence mostly foreign products. Therefore, security controllability of China's industrial internet is an arduous task. The organization has monitored 8.2 million detections of fingerprints on industrial control equipment connected into the Internet and found 465,000 IP addresses from 160 countries had made fingerprint detection on China's

[1]Source: CNCERT.

industrial control equipment connected into the Internet, the number seeing an explosive growth in comparison with that in 2016.

3. **Data breach happens frequently.**

With the crossover development of the Internet, data and information is available on the web, mobile phones and cloud. Driven by interests of black industrial chains of the Internet, website data and personal information breach happens frequently, with the scale and scope expanding, posing threat to personal property and safety. At the end of 2014, China began to see continuous data breaches. In December 2014, 130,000 users' ID numbers and passwords were leaked out from 12306.cn, the official website of Ministry of Railways; in February 2015, customer order information of famous chain hotels like Orange Hotel, Jinjiang Inn and high-end hotels like Marriott and Sheraton was let out; in August 2015, account passwords of over six million users of damai.cn, the largest ticketing website in Chinese, were revealed; in October 2015, sensitive information of over 100 million users of NetEase mail was disclosed; in April 2016, information of 200,000 children was sold in package; in August 2016, data breach took the life of Xu Yuyu, who had been admitted by a university; in July 2017, the users' shared information on Baidu Netdisk was leaked out.

Column 20 Xu Yuyu's Death Caused by Telecommunication Fraud

On August 19th, 2016, Xu Yuyu, a 18-year-old girl from a poor family in Linyi, Shandong Province, who had just got the admission notice of the university, received a fraud call in which the callers pretended to be clerks of the local education and finance bureaus, telling her that they would grant her RMB2, 600 *yuan* of subsidies for her university study. She believed them and did what they told her to do next. Finally, she sent RMB9, 900 *yuan*, which had been borrowed by her parents with great difficulties, to the cheaters. When she found herself cheated, she was heart-broken. On her way back from the police station where she reported the case, she died of heart attack. Her death attracted the attention of the whole society.

The police attached importance to the case. They set up a task group for investigation. On July 19th, 2017, the first-trial sentence of the case was announced. Chen Wenhui, the principal criminal, was sentenced to life imprisonment for fraud and illegal obtaining of personal information. All his property would be confiscated. Other six criminals were sentenced to 15 years to 3 years of imprisonment and were all fined. On September 15th of the same year, the second trial upheld the verdict of the first trial. In that case, the hacker Du Tianyu was sentenced to six years of imprisonment and fined RMB 60,000 *yuan* for illegal obtaining the citizens' personal information.

4. **Security of the domain name system has been improved.**

The domain name system (DNS) used to be a weak aspect affecting the operation of the Internet in China, for it was faced with denial of service attacks (DDoS attacks) and frequent resolution distortion attacks on important websites. In 2014, there occurred large-scale DNS breakdowns, with many web portals unable to be opened. According to IP positioning, those DNS attacks were launched from overseas. With the enhancement of key domain name security protection, the security of China's DNS has improved. In recent years, DDoS attacks on key DNSs have not caused serious damage to domain name resolution service of the systems and few attacks affecting resolution success rate have been found in basic telecommunication businesses. DDoS attacks on global root domain name servers and their mirroring in June 2016 were tackled in time and thus left no harm on China's DNS security.

Column 21 Large-scale DNS Resolution Breakdown in China in 2014
At about 15.10, January 21st, 2014, there occurred abnormality in the root server resolution of China's general top DNSs, with some domestic users unable to visit the websites of .com. Baidu, Sina, Tencent, JD.com and Youku and some other websites could not be visited intermittently. Many websites were found to be resolved to an American IP address 65.49.2.178, which could not be visited at all. According to primary statistics, two thirds of the websites in China fell victim to the attack. The resolution of root servers throughout the country was recovered at 16 o'clock that day.

5. **There are an increasing number of problems concerning the mobile Internet security.**

There are more and more attacks on the mobile Internet, as it is more and more frequently used, so it has become a new focus to protect the security of the mobile Internet. In the first three quarters of 2017, nearly 1.68 million malicious programs attacking the mobile Internet were captured through autonomous monitoring and sharing. The number keeps increasing in recent seven years. The top two types of the malicious program are rogue software (54.5%) and money wasting software (37.7%). In the first three quarters of 2017, there were 1.46 million download chains of malicious programs attacking the mobile Internet, nearly 1.2 times of the number in the whole year of 2016. The programs were transmitted 20 million times in the three quarters. Malicious APPs have been controlled effectively. By September 2017, CNCERT had informed 141 registered APP shops and advertising websites of net disks and cloud disks, of 8,166 malicious APPs incidents. Notably, the security of mobile payment cannot be neglected. Social media account theft, SMS Trojan link and security code cheating are the most frequently-used cheating

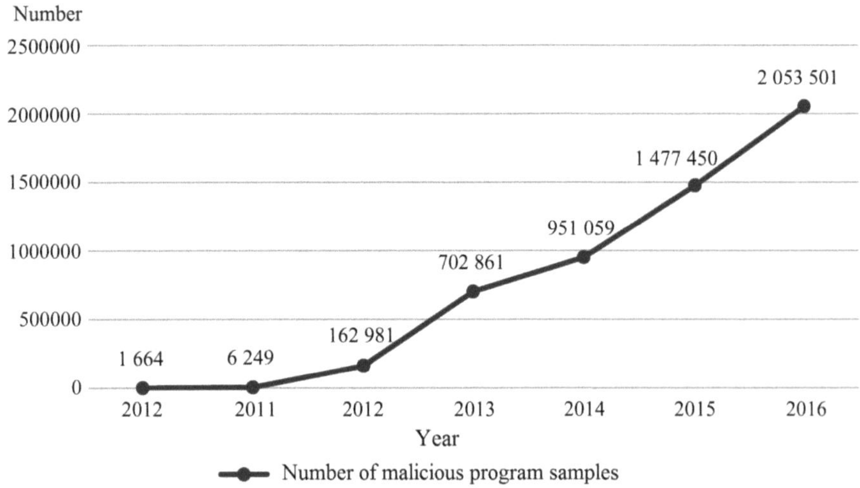

Fig. 6.2 Number of mobile internet-oriented malicious program samples (2010–2016) (*Source* CNCERT)

means.[2] The number of mobile Internet-oriented malicious program samples is shown in Fig. 6.2.

6. **There are frequent problems with security of IoT smart devices.**

Terminal devices and network devices like smart wearables, house appliance and routers are becoming important components of people's work and life, but there are an increasing number of cyber attacks on IoT smart devices. Attackers can get the control authority by taking advantage of the vulnerabilities of the smart devices to make the IoT into a large-scale botnet or steal the users' data and hijack network traffic for transaction on the underground market of hackers. In the first three quarters of 2017, CNVD recorded 531 vulnerabilities of IoT smart devices, involving web cameras, smart routers, intelligent house appliance and intelligent gateways. The vulnerabilities were permission bypass, information breach and command execution. The weak password (or built-in default password) vulnerabilities can easily be used for malicious code attacks.[3] With the popularization of drones, autonomous vehicles and smart appliances, as well as the development of smart cities, there will be more frequent attacks on IoT smart devices.

7. **There is an increasing number of cloud security incidents.**

With the accumulation of cloud services and data, there are more and more APT and 0Day attacks on cloud associated with vulnerabilities, viruses and unknown threats. At the end of 2014, a famous game corporation on Alibaba Cloud suffered a

[2]Source: China UnionPay, *Analysis Report on Mobile Payment Security Survey (2016).*

[3]Source: CNCERT.

DDoS attack, the biggest of its kind in the world at that time. The attack lasted as long as 14 h and its peak value traffic reached 453.8 Gbps.[4] In the first half of 2015, several thousand host computers of Tencent Cloud suffered over 10,000 DDoS attacks by external hackers, with the highest number of 2,600 within one month; the biggest attacked traffic was increasing month by month, with the peak value at nearly 300 Gbps in one month.[5] Cloud Security Alliance summarized the "12 cloud security threats" as data breach, credential theft and useless identity certificate, interface and API being hacked, system vulnerability being used, account hijack, malicious insiders, APT parasite, permanent data loss, insufficient investigation, cloud service abuses, DoS attacks and sharing technology.[6]

6.1.2 There Are More Means of Cyberattacks

1. **China is becoming a major victim of APT attacks.**

Today, frequent APT attacks are giving rise to intelligence of governmental organizations and businesses being stolen, industrial systems being destroyed and the financial system suffering losses, which threatens the national security and citizens' rights and interests. According to accumulated statistics of 360 Threat Intelligence Center, by the end of December 2016, 36 APT organizations from home and abroad attacking China and nearly 200 domestic suspected targeted organizations of APT attacks had been detected. China is becoming the first targeted country of APT attacks. In the single year of 2016, APT attacks affected at least over 10,000 computers within the country in 31 provincial regions, covering government, infrastructure, education, research and large businesses.[7] For example, White Elephant Attack and Manlinghua Attack have all been APT attack incidents targeting the information systems of China. At present, more networking production systems and infrastructure are faced with tremendous security threats of APT attacks.

2. **DDoS attacks happen frequently.**

DDoS attacks have become the most attractive and popular among hackers. According to NSFOCUS's science and technology monitoring statistics, in comparison with 2016, the first half of 2017 saw fewer attacks. Altogether, about 100,000 DDoS attacks were detected, an increase of 30% in comparison of the number in the second half of 2016; the total attacked traffic was 16,000 Tbps, a

[4]Source: Alibaba Cloud's Declaration on Alibaba Cloud Games' Users Being Attacked by Hackers, December 24th, 2014.

[5]Source: Tencent Cloud, *White Paper of Tencent Cloud Security*.

[6]Source: *International Cloud Security Trend and Practice*, a keynote speech by Cloud Security Alliance at Cloud + Future Summit hosted by Tencent.

[7]Source: 360 Internet Security Center, *Report on China's Internet Security (2016)*.

decrease of 38.4% in comparison with that in the second half of 2016. On the other hand, the number of attacks on extra-large traffic exceeding 300 Gbps is increasing.[8] Government, finance and the game industry are seriously afflicted areas, and even all sectors are faced with the threat from DDoS attacks. Besides, the means that hackers adopt are becoming smarter and smarter. While launching DDoS attacks, hackers begin to steal data, infect virus and cheat to obtain greater benefits.

3. **Trojan and botnet infection is becoming less serious.**

According to CNCERT sampling monitor, in the first three quarters of 2017, the incidents in which Trojan and Botnet controlled host computers by controlling the servers involved 16.62 million IP addresses of host computers, including 11.385 million domestic addresses and 5.235 million overseas ones. In 2016, the three numbers were respectively 25.84 million, 17 million and 8.85 million. Obviously, they all decreased. In the first three quarters of 2017, there were 130,000 domestic IP addresses of servers and 52,000 overseas ones controlled by Trojan or botnet. The numbers increased in comparison with those in 2016.[9]

4. **Ransom virus is witnessing a concentrated outbreak.**

Ransom virus is a new type of computer virus, threatening the safety of local data and smart devices. According to the detection of CNCERT, in the first three quarters of 2017, 139,300 samples of malicious programs with Ransom virus were found at traditional PC terminals. The target of ransom virus attacks is shifting from personal terminal devices to business users'. The attacks are aimed at encrypted database of businesses. For personal terminal devices, traditional PC extortion is done through the encrypted database and mobile terminal extortion is done through the encrypted devices. In May 2017, WannaCry, a kind of ransom virus, swept the world, and the intranet of many businesses as well as education, medical care, power, energy and banking and other sectors were all affected. At some service stations of PetroChina, the online payment by rechargeable cards and bank cards and third-party payment could not be done.

Column 22 Attack Means

APT is the acronym of Advanced Persistent Threat. It is a cyber attack and invasion aiming to steal core data. In APT attacks, the hacker, by hiding him/herself, steals the data from the target regularly in a planned and organized way. It is cyber-espionage since the hacker steals data and collects intelligence in the cyberspace.

DDoS refers to Distributed Denial of Service. Using this means, the hacker takes more than one computer as the attacked platform by taking advantage of the client/server technology. He/she launches DDoS attacks on one or more

[8]Source: NSFOCUS, *Report on DDoS and Web Application Attacks* (the first half of 2017).
[9]Source: CNCERT.

targets. The master control program can activate within seconds hundreds of thousands of runnings of the agent program, and thus increase exponentially the power of DDoS attacks.

Botnet refers to the computer group subject to the hacker's concentrated control. The hacker can make the host computer infected with Trojan or botnet by taking the same malicious action through the one-to-many command and control channel. For example, the hacker can launch a DDoS attack on the targeted website or send a large number of junk mails at the same time. attacks.

Ransom virus is a new type of computer virus, spread through e-mails, Trojan and websites' malicious code. The hacker encrypts the files through all kinds of encryption algorithm, so that the user of the infectious computer cannot decrypt until he/she gets the secret key.

6.2 The Cyber Security System (Mechanism) Is Being Built and Improved Fast

Since 2012, guided by the cyber security strategic plan and based on the establishment of cyber security monitoring and emergency system, cyber security assurance capacity of China has been improved through the facilitation of the cyber security system and mechanism. The country continues to be aware of the present status of cyber security administration and the focus of cyber security assurance, constantly improve the relevant laws and regulations as well as working mechanism, and cultivate a number of professional supporting organizations.

6.2.1 Guidance of the Strategic Plans on Cyber Security

Faced with the increasingly complicated cyber security, China persists in planning first. Since the 18th CPC National Congress, the country has launched a series of strategic plans concerning cyber security, including the *National Cyberspace Security Strategy*, *Strategic Outline for the National Informatization* and *National Informatization Planning for the Thirteenth-five-year Plan Period*, which provide guidance for the cyber security administration. In July 2016, the *Strategic Outline for the National Informatization* was released, which proposes that a correct philosophy on cyber security should be formed. According to the *Outline* we should prevent and handle cyber security threats actively, enhance our capacity of preventing and deterring any threats to cyber security, safeguard our cyber sovereignty and national security, and strengthen the critical information infrastructure security protection and basic work in cyber security. In December 2016, Cyber Administration of China released the *National Cyberspace Security Strategy* as the guiding document for the national cyber security administration. Guided by the general national security concept, the document focuses on development and security, the two major issues, to promote peace, security, openness, cooperation and orderliness of the cyberspace, clarifies the four principles and nine tasks for China's cyber security. In December 2016, the State Council issued the *National Informatization Planning for the Thirteenth-five-year Plan Period*, which also focuses on both security and development. This document proposes the important task to improve the cyber security assurance system as well as other tasks and projects like "enhancing the top design for cyber security", "constructing the security assurance system for critical information infrastructure", "all-day and all-dimensional perception of cyber security" and "enhancing innovative capacity concerning cyber security".

6.2.2 *The Law and Regulation System Concerning Cyber Security Keeps Being Improved*

1. **Laws of cyber security are made and improved.**

Cybersecurity Law of the People's Republic of China was put into enforcement on June 1st, 2017, stipulating the basic principles, tasks and guidelines for the national cyber security administration and defining the rights, obligations and responsibilities of governmental agencies, businesses, social organizations and individuals. It is a basic and comprehensive framework law concerning cyber security. In July 2015, *National Security Law of the People's Republic of China* was put into enforcement, proposing to build the cyber and information security assurance system, improve the cyber and information security protection capability, enhance the innovative research and development application of network and information technology, and defend the national cyberspace sovereignty, security and development interests. *Amendment (IX) to the Criminal Law of the People's Republic of China* explicitly defines the cybercrime of fabricating and spreading false information about dangers, epidemic, disasters or police. In January 2016, *Counterterrorism Law of the People's Republic of China* was put into effect, stipulating terrorism countering in cyberspace.

2. **Policies and regulations concerning cyber security have been launched.**

In 2003, *Opinions of the Leading Group of the National Informatization on Enhancing Information Security Assurance* was issued, which defined for the first time the basic principles for cyber security and proposed the guidelines and main tasks of enhancing cyber security assurance. Later, a series regulations were formulated, including *Measures for the Administration of IP Address Archiving*, *Administrative Measures for the Multi-Level Protection of Information Security*, *Measures for the Administration of Criptography for Electronic Certification Services*, *Opinions on Promoting Informatization and Ensuring Information Security*, *Implementation Measures for Network Security Information Notification* and *Measures for the Administration of Communication Network Security Protection*. In 2012, the Standing Committee of the National People's Congress launched *Decision of Standing Committee of the National People's Congress on Strengthening Protection on Internet Information Content*. In July 2013, the Ministry of Industry and Information Technology released *Provisions on Protecting the Personal Information of Telecommunications and Internet Users* to regulate the collection and use of users' information in the telecommunication and Internet information service and to protect the legal rights and interests of telecommunication and Internet users. In May 2014, Office of the Central Leading Group for Cyberspace Affairs issued *Notice of Enhancing the Security Management of CPC and Governmental Websites* to improve the cyber security assurance and administration capacity of CPC and governmental organizations to ensure safe and healthy development of their websites. In August 2014, the Ministry of Industry and Information Technology released *Guiding Opinions on Enhancing the Network Security of the Telecommunication and Internet Industries*, proposing to strengthen the cyber security surveillance in network infrastructure security protection, cyber security emergency response, critical security-controllable soft and hardware applications and network data and individual users' information protection. In January 2017, Office of the Central Leading Group for Cyberspace Affairs issued the *National Emergency Response Plan for Cyber Security Incidents*. In August of the same year, Ministry of Industry and Information Technology modified *Measures for the Administration of Internet Domain Names* to optimize the registration administration concerning IP addresses, domain names and websites.

6.2.3 The Working Mechanism for Cyber Security Is Being Improved

1. **The cyber security review system is being established.**

In May 2014, competent authorities proposed to launch cyber security review, which would cover information technology products and services used by systems concerning national security and public interests. In December 2014, Office of the Central Leading Group for Cyberspace Affairs launched *Opinions on Enhancing the Network Security of Cloud Computing Service of CPC and Governmental Sectors*, proposing to launch security reviewing covering cloud computing service for CPC and governmental organizations. The review would be focused on the security and controllability of cloud computing service. In May 2017, Cyber Administration of China launched *Measures for the Security Review of Network Products and Services* (for Trial Implementation) to set up and improve cyber security review and clarify the necessity of review for key network products related to national security. So far, Jinan E-governance Cloud Platform by Inspur, Chengdu E-governance Cloud Platform (II) by Sugon and Alibaba Cloud E-governance Platform by Alibaba Cloud and some others have passed the cyber security review.

2. **Inspection is going on for critical information infrastructure network security.**

General Secretary Xi Jinping requires that cyber security inspection should be enhanced so that we can know everything including the risks and loopholes about cyberspace. According to Xi, the inspection result should be published and problems be solved in time. To meet this requirement, since 2014, Office of the Central Leading Group for Cyberspace Affairs have carried out major cyber security inspections on network and information systems of central CPC departments and governmental ministries, the people's organizations and key industries. In 2016, cyberspace administrations of all provinces (autonomous regions and municipalities directly under the Central Government), with Office of the Central Leading Group for Cyberspace Affairs as the coordinator, launched the first nationwide inspection of critical information infrastructure security. The whole country was taken as a whole in the inspection, which enhanced the administration, prevention, rectification and construction, providing data and reference. In that inspection, more than 54,000 information systems of 31 provinces (autonomous regions and municipalities) were involved, covering 90% of regions above the county-level.

3. **Multi-level protection of information security is continuously carried out.**

In January 2007, Ministry of Public Security released *Administrative Measures for the Multi-Level Protection of Information Security*, which stipulates that multi-level protection should be done for information and information carriers in accordance with the importance of the information and its carriers and that a series of working mechanisms of multi-level protection of information security from the Central Government to local governments should be worked out. The multi-level protection involves grading, registration, security construction and rectification, information security grading evaluation and information security inspection. Security is classified into five levels in accordance with the degree of the damage of and threat to the information, and the security protection capacity varies with the different level of the information system security. According to *Cybersecurity Law of the People's Republic of China*, the country pursues multi-level protection of cyber security, and network operators are supposed to fulfill their security protection obligations, protect their networks from disturbance, damage or unauthorized visits and prevent data from being breached, stolen or tampered.

6.2.4 There Is Strong Support from Cyber Security Assurance Service Organizations

1. **National assurance service organizations are continuously established and improved.**

After two decades of development, six national assurance organizations for network and information security emergency technologies have been founded in China. (1) National Virus Emergency Response Center (CVERC). It is the only virus emergency response center of China, made up of China Internet Virus and Attack

Monitoring and Warning Center and National Mobile Internet Application Security Management Center, responsible for quickly finding and disposing of viruses and attacks. CVERC has the most authoritative malicious code sampling database storing 15 million virus samples. (2) China Information Technology Security Evaluation Center. It is an organization engaged in information technology security testing and risk assessment, with laboratories of basic vulnerabilities research, application software and product security testing, system hazard analysis and evaluation equipment R&D. (3) CNCERT. Relying on the Internet security monitoring platform, it is responsible for prevention, spotting, warning and coordinated disposal of cyber security incidents, having set up branches in 31 provinces (autonomous regions and municipalities), and a stable cyber security reporting system with 580 member organizations. (4) National Information Notification Center of Network and Information Security. It is the national network and information security notification outlet, reporting any information concerning cyber security. (5) National Research Center for Information Technology Security. It is engaged in security analysis and research of information technology products and systems, assurance of the basic networks and key information systems, and R&D of information security technologies with independent intellectual property right. (6) China Information Security Certification Center. It is responsible for information security certification in accordance with national laws and regulations on mandatory product certification and information security administration.

2. **National social organizations of cyber security have been set up.**

With the rise of the strategic position of cyber security, a number of national social organizations of cyber security like associations and alliances have been set up. In March 2016, Cybersecurity Association of China was founded in Beijing, involving industrial, educational, research and application institutions, businesses and individuals engaged in cyber security. It has 260 member organizations and 321 individuals, covering major Internet and cyber security businesses, authoritative research institutions and cyber security experts. They have launched with local cyberspace administrations a series of activities like Cybersecurity: China Tour, to improve cyber security awareness and skills. In December 2015, China Cybersecurity Industry Alliance was founded in Beijing, involving over 140 representative businesses of cyber security, such as Bluedon, NSFOCUS, Venustech and Westone. Cloud Security Alliance was jointly founded in January 2016 by Tencent Cloud, AsiaInfo, IBM, Symantec, Topsec and Sangfor.

6.3 Breakthroughs Have Been Made Rapidly in Cyber Security Industry and Technology

Cyber security has become a "major security" that all circles in society are faced with. After decades of development, breakthroughs have been made in independent and controllable R&D concerning cyber security technology, which is witnessing new development opportunities. Relevant standards and talent cultivation systems are being established.

6.3.1 Breakthroughs Have Been Made in Cyber Security Technology

China is enhancing the establishment of new technology research and security standards and carrying out the R&D of a series of security product and system platforms by overcoming a number of key technical difficulties. The country has developed independently ZUC Algorithm that can match mainstream international cryptographic algorithm. It has achieved progress in joint development of secure and reliable basic soft and hardware. Innovative breakthroughs have been made in domestic chip R&D, such as Loongson, SW, Phytium, Hisilicon and Spreadtrum. Soft operation systems developed independently are not only "nearly usable", but "usable". Databases developed independently are widely used in governmental departments and key areas. Complete computers made in China based on basic soft and hardware made in China have been put into trial use in some CPC and governmental departments. Security prevention and protection systems of industrial control are being industrialized.

1. **Basic technologies have been developed one after one.**

For a long time, China had to import chips, servers, databases and operation systems. Since the 18th CPC National Congress, the situation has changed. China Information Terminal OS Alliance was founded in 2013 to build a unified ecosystem and a new market for domestically-made operation systems. Operation systems of iSoft, NeoKylin and Syswin were released in 2014. In 2016, Loongson announced that 3A 3000 quad-core processer saw the completion of the tape-out and the system test, Huawei released SOC Hisilicon 960, Rockchips released high-performance and high-scalability universal flag chip RK 3399, and Phytium released FT 2000. Meanwhile, breakthroughs have been made in China in key chip-manufacturing technologies, covering 5 nm plasma etcher, high-purity metal sputtering target and lapping liquid. Dramatic development is seen in the IC industry. In 2017, China succeeded in developing the photon quantum porotype computer of 10 photon quantum entanglement manipulation, breaking the world record, which marks that the country is a leader in quantum computing.

2. **Security protection technology has developed from passive defense technology to active defense technology.**

The Chinese Government has appropriated fund for the establishment a number of national information security R&D centers concerning the security of industrial control, cloud and e-government. Thus R&D of high-performance security gateway, UTM, SOC and firewall has been supported, covering cloud computing, IoT, mobile Internet and next-generation Internet, to ensure the security of the information system in e-government, telecommunication, finance and transportation. In 2016, manufacturers related to cyber security kept improving the performance of their products to meet the security demand of different sectors. In February that year, Hillstone, a Chinese manufacturer, won the Recommended Award with its

99% comprehensive threat detection rate and No.1 total ownership cost in the 2015 next-generation firewall test and selection result published by NSS labs, a renowned independent international security research and testing institution. In April, KUANGN launched the first portable system platform for industrial control security threat evaluation; in May, NSFOCUS released IDS-ICS, a totally new industrial control invasion detection system, honored as China's first industrial control invasion detection product. In November, "Theory and Core Method of Cyber Mimic Defense" passed verification as China's key innovative theory and method in cyber defense. Besides, it has become an important way of countering targeted attacks to use big data technology to analyze behaviors, sense the situation and trace the origin. Cutting-edge technologies like deep learning, AI, block chain and quantum computing are emerging. The application of their integration with cyber security is being studied and developed.

6.3.2 The Cyber Security Industry Is Developing Rapidly

1. **The industrial scale is increasing dramatically.**

According to the prediction CCID, China's cyber security industrial scale from 2013 to 2017 would increase from RMB53.65 billion *yuan* to RMB 106.64 billion, with a compound annual growth rate of about 25%. Promoted by policies, it will witness its golden development period. It was expected that in 2017, the scale would be RMB13.554 billion *yuan*, with an increase of over 27.1%. The PC security market is dominated by domestically made brands and over 99% of domestic PC users enjoy using free security software.[10] With the rapid increase of cyber security industrial scale, there have emerged a number of leading businesses in that field. By the end of 2016, there had been over 1,000 cyber security businesses, with over 30 of them having the annual revenue of over RMB100 million *yuan*. A large number of leading traditional businesses in cyber security have emerged, including Venustech, Sangfor, NSFOCUS, 360 Enterprise Security Group, Topsec, Westone and H3C, and a number of emerging and startup businesses, including Guominrenzheng, Qingteng, Jowto, Safedog, Winicssec and Bangcle. Large Internet businesses like Alibaba, Tencent and Baidu have extended their business to cyber security and hence brought new energy into the industry. By June 2017, 13 businesses, including Venustech, Bluedon and Westone, had been listed with the A-share stock exchange, and 38 businesses, including Uxsino, TIPS and SBR-info, had been listed with the New OTC Market.[11] In general, despite the increase of the strength of its cyber security businesses, China still has a long way to go to meet the cyber security demand and to catch up with those countries with stronger cyber security businesses. The industrial scale and growth of China's cyber

[10]Source: CCID, *Blue Paper of China's Cyber Security Development (2016–2017).*

[11]Source: Aqniu.com, *China's Top 50 Cyber Security Business (the first half of 2017).*

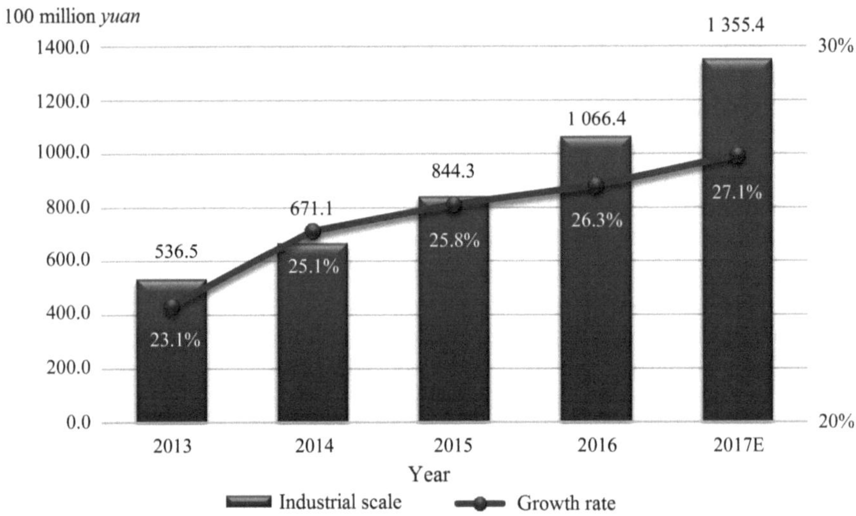

Fig. 6.3 Industrial scale and growth rate of China's cyber security industry (*Source* CCID)

security industry is shown in Fig. 6.3 and cyber security businesses of the country are listed in Table 6.1.

2. **The industrial structure of cyber security is being improved.**

The cyber security industry is made up of five parts, namely, basic security, IT security, disaster recovery and backup, credible cyberspace identity service and others. In the industrial structure of cyber security of China in 2016 (See Fig. 6.4), IT security was the dominant part, accounting for 48.4%; basic security, disaster recovery and backup, and credible cyberspace identity service accounted respectively for 5.6%, 13.1% and 32.8%. According to IDC statistics, in the IT security industry of 2016, IT security hardware accounted for 50.5%, and IT software and service saw dramatic increase, accounting for 10.7%.[12] Cyber security products developed by China on its own have covered physical security, communication security, data security, application security, security management platforms and security of new technologies and applications. In the country, complete cyber security industrial structure has been formed, involving security chips, network and border security products, data security products, application security products and security service. China's IT security industrial structure from 2013 to 2016 is shown in Fig. 6.5.

3. **The industrial chain of cyber security has basically been formed.**

At present, the upstream and downstream industrial chain of cyber security is made up of academic institutions like universities, research institutes and national

[12]Source: CCID, *Blue Paper of China's Cyber Security Development (2016–2017).*

Table 6.1 China's cyber security businesses

No.	Business	Key business areas	Revenue in 2016 (RMB 100 million yuan)	Headquarter location
1	Venustech	Firewalls & NGFW, GAP, IDS/IPS, UTM, database security, DLP, VPN, vulnerability scanning, SOC & NGSOC, assessment and reinforcement, and security operation and maintenance	19.27	Beijing
2	Sangfor	Firewall & NGFW, UTM, Internet access management, VPN, and mobile terminal security	17.5	Shenzhen
3	NSFOCUS	Firewall & NGFW, IDS/IPS, UTM, host protection and self-adaption, anti-DDoS equipment, database security, vulnerability scanning, Web application scanning and monitoring, WAF, security consulting service, assessment and reinforcement, and security operation and maintenance	10.89	Beijing
4	360 enterprise security group	Firewall & NGFW, GAP, UTM, terminal safeguard & virus prevention, EDR, VPN, code audit, WAF, mobile app security, mobile terminal security, threat intelligence, big data security, APT, SOC & NGSOC, and penetration testing service	–	Beijing
5	AsiaInfo	UTM, host protection and self-adaption, terminal safeguard & virus prevention, EDR, spam prevention, cloud infrastructure security, mobile terminal security, APT, phishing prevention, and SOC & NGSOC	–	Chengdu
6	Topsec	Firewall & NGFW, IDS/IPS, DLP, VPN,cloud infrastructure security, and big data security	–	Beijing
7	Westone	Firewall & NGFW, IDS/IPS, VPN, disk encryption, file security, and encryptor	17.99	Chengdu
8	H3C	Firewall & NGFW, IDS/ IPS, UTM, VPN	260	Hangzhou
9	DBAPP security	database security, Web application scanning and monitoring, WAF, bastion host/operation and maintenance security, big data security and other safeguard tools	–	Hangzhou
10	Meiya Pico	big data security, security and forensics, public opinion monitoring, safety education training, and security integration service	5.92	Xiamen

Source Sales revenue of H3C

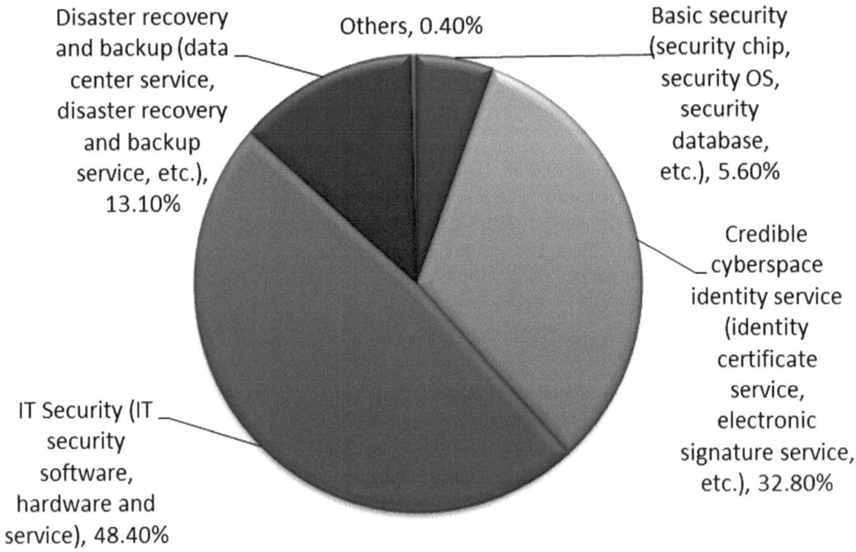

Fig. 6.4 China's cyber security industrial structure (2016) (*Source* CCID)

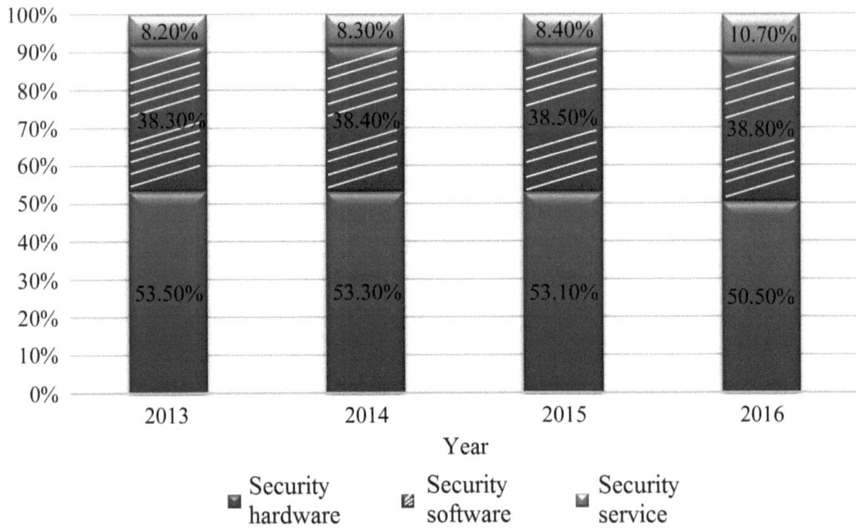

Fig. 6.5 China's IT security industrial structure (2013–2016) (*Source* CCID)

laboratories, basic software and hardware manufacturers, platform data integration manufacturers and terminal users (See Fig. 6.6). Higher-education institutions belong to the upstream, responsible for basic theoretical research and core technological breakthroughs concerning cyber security; basic software and hardware manufacturers belong to the middle-stream, responsible for the industrialization of

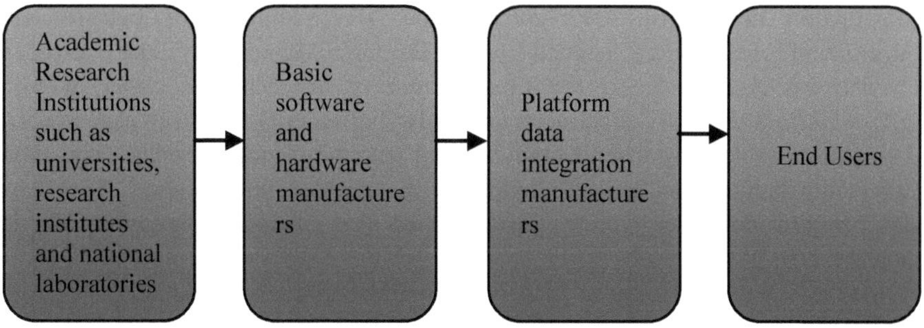

Fig. 6.6 Industrial Chain of cyber security (*Source* CCID)

the research results, manufacturing basic software and hardware of cyber security; platform data integration manufacturers and terminal users belong to the downstream, where the integration manufacturers are responsible for second development of the basic software and hardware into terminal products of different functions for terminal users.

4. **There are more industrial parks of cyber security.**

To seize the opportunity of the industrial development concerning cyber security, many provinces, relying on their industrial foundation and elements, keep speeding up the industrial layout of cyber security and have built their industrial parks in that field. Since 2000, Ministry of Science and Technology has approved three national industrial bases for achievement industrialization of information security. They are the Sichuan, Hubei and Shanghai bases. In April 2016, cooperating with China Electronics Technology Group Corporation, Sichuan Province invested RMB13 billion *yuan* in constructing Chengdu Cyber Security Industrial Park of China Electronics Technology Group Corporation to make it into a national industrial park of network information security. The Park relies on the cooperation between the military and civil forces, making use of the advantages of China Cyber Security–a subsidiary group of China Electronics Technology Group Corporation–in cryptography, information security and physical security. It focuses on information security, secure information systems, safe operation, security system integration and security service and testing. In September 2016, the Yantai Municipal Government, in cooperation with China Electronics, begun to construct the National Cyber Security Industrial Park (Yantai), with R&D and Industrialization Center of the Sixth Research Institution of CEC and the business incubators as the core. The first projects have been allowed into the park, covering CRS, microwave communication electronics, information security prevention and protection product R&D, and industrialization. Wuhan of Hubei Province has begun to construct national cyber security talent and innovation bases, including a cyber security innovation base, a public incubator center, a cyber security research institute, a cyber security talents training center, a cyber security academy, a talent community, a super-computing center and a sharing center. The National Cyber Security Academy was put into construction in August 2017. In October 2017, Ministry of Industry and Information Technology signed with Beijing Municipality the *Strategic Cooperation Agreement on the Construction of the National Cyber Security Industrial Park* to jointly construct a national cyber security industrial park to boost the industry to develop toward high-end and autonomization. Besides, Hangzhou, Hefei and Changsha are active in laying out the cyber security industry. They have also begun to construct similar industrial parks and enhance their support for the industry.

6.3.3 Cyber Security Standardization Is Speeding up

The cyber security standard is an important technical support for the construction of the cyber security assurance system and a key tool for safeguarding the national interests and security. Having witnessed three development stages, namely, infancy, starting and rapid development, China's cyber security standard system has been basically constructed, laying a foundation for the normalization and sustaining development of the standardization. In April 2002, National Information Security Standardization Technical Committee (NISSTC for short, with its number as TC260) was officially founded, fully responsible for planning and managing the national standards of cyber security. By November 2017, it had released 207 relevant standards, with the number having increased. China has participated in the formulation of a number of international standards and has achieved a lot in cryptography algorithm, entity authentication, information security incident administration, cloud computing and virtualization, big data security, reliable supply chain, and biometrics identification. The number of national standards released by National Information Security Standardization Technical Committee in the years from 1995 to 2016 is shown in Fig. 6.7.

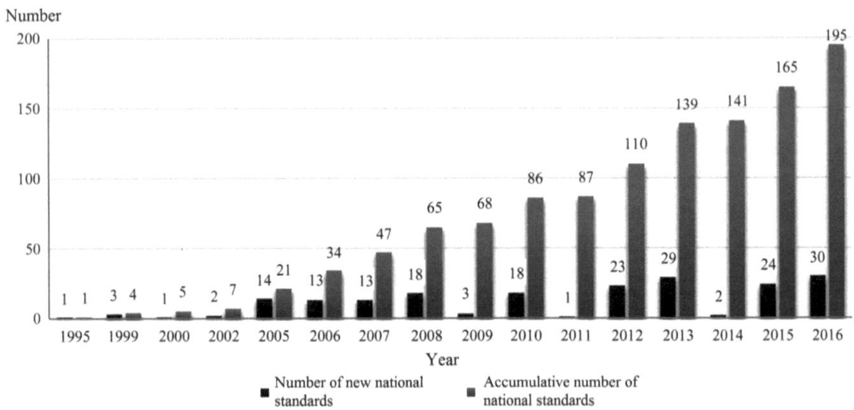

Fig. 6.7 Number of national standards released by national information security standardization technical committee in the years from 1995–2016 (*Source* CCID)

Column 23 Structure and Function of NISSTC

National Information Security Standardization Technical Committee (NISSTC) is in charge of organizing information security standard formulation of China. It has 8 working groups responsible for the standardization of technology, mechanism, service, administration and assessment of information security[13] (See Fig. 6.8). The structure is as follows:

The Working Group of Information Security Standard System and Coordination is responsible for studying the information security standard system; following up the international information security standard development; studying and analyzing the demand for domestic information security standard application; and studying and proposing new projects and suggestions. **The Working Group of the Confidentiality Standards for Classified Information Systems** is responsible for studying and proposing new projects and suggestions, and formulating and modifying the confidentiality standards for classified information systems. **The Cryptography Technology Working Group** is responsible for studying and formulating cryptography algorithm, cryptography modules and key management standards. **The Identification and Authorization Working Group** is responsible for analyzing, studying and formulating domestic and foreign PKI/PMI standards. **The Information Security Assessment Working Group** is responsible for surveying and studying the domestic and foreign testing standards and their trend and proposing testing standard projects and relevant plans. **The Communication Security Standard Working Group** is responsible for surveying and studying communication security standards and their trend, proposing the communication security standard systems, and formulating and modifying communication security standards. **The Information Security Management Working Group** is responsible for researching into information security management standard systems and formulating relevant standards. **The Special Working Group of Big Data Security Standards** is responsible for formulating the security standards relevant to big data and cloud computing.

1. *Information Security Standard System* proposes the blueprint for standardization information security.

In collaboration with Ministry of Public Security, Ministry of State Security, State Secrets Bureau, State Cryptography Management Commission, Certification and Accreditation Administration and Headquarters of the General Staff of PLA, NISSTC categorized and sorted the existing information security standards, and by following up the development of international information security standards and

[13]Source: www.tc260.org.cn/.

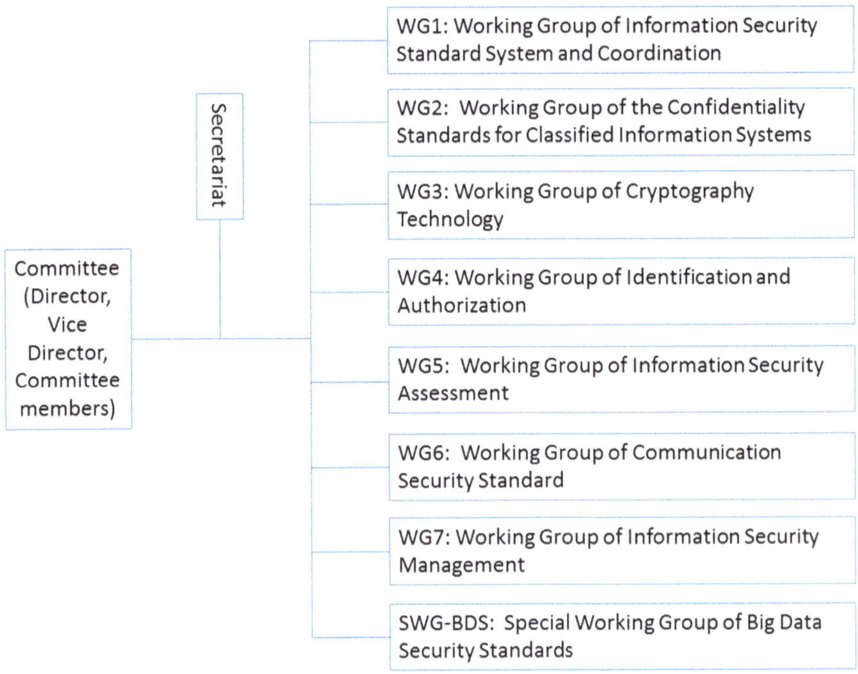

Fig. 6.8 Structure of national information security standardization technical committee

the domestic demand for information security standards, proposed a systematic, comprehensive and reasonably distributed information security standard system, which, after being modified for times, has been formed into the *Information Security Standard System* (Version 2.0). This system contains the standard system framework and tables, with information security standards categorized into seven kinds: basic, technology and mechanism, management, evaluation, cryptography, confidentiality and communication (See Fig. 6.9). The information security system is an organic integration made up of interrelated standards within the information security field and the blueprint for existing, should-exist and to-be-formulated information security standards.

2. **Cyber security standardization is progressing.**

On the basis of the framework of the *Information Security Standard System*, China, in recent years, has been paying attention to the formulation of standards concerning cyber security and has been promoting the standardization. In August 2016, Office of the Central Leading Group for Cyberspace Affairs, General Administration of Quality Supervision, Inspection and Quarantine, and Standardization Administration jointly issued *Several Opinions on Strengthening National Cybersecurity Standardization*, in which lay-out is made for China's cyber security standardization. The document requires to establish the standard sharing

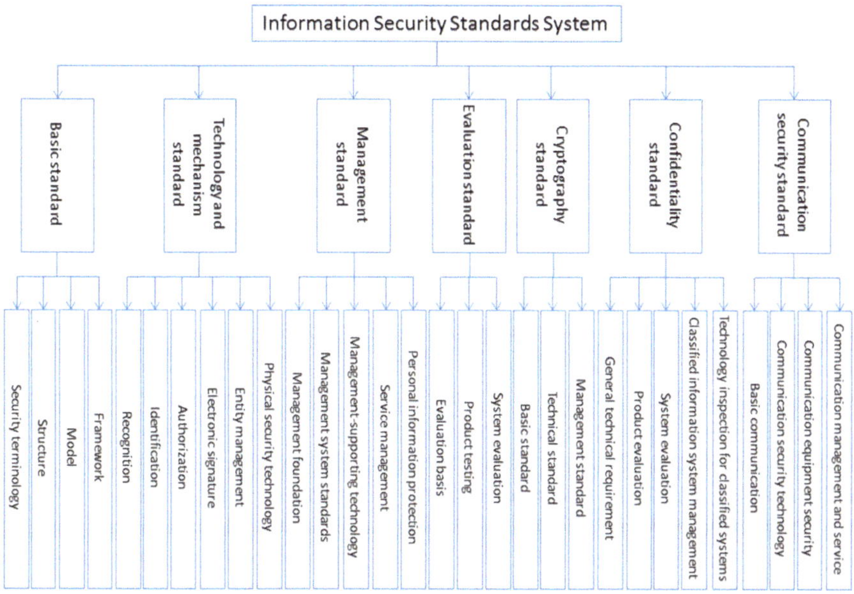

Fig. 6.9 General framework of the information security standard system (*Source* Xu Yuna and Luo Fengying, Analysis of China's Information Security Standard System, Confidentiality Science and Technology, 2014(1):17–20)

mechanism for key projects and important science and technology projects and enhance the transfer of the results of key projects or research projects into national standards; to pursue the principle of "the priority of the urgent", that is to say, to first formulate the standards in need, for instance, the standards for critical information infrastructure protection, cyber security review, big data security, person information protection, next-generation communication network security, Internet TV terminal product security and cyber security information sharing.

6.3.4 The System of Cyber Security Talents Cultivation Is Being Improved

Talents are the foremost resources for cyber security. The competition for cyberspace is fundamentally the competition for talents, so it is necessary to build an excellent talent team to become a strong country in terms of cyberspace. According to statistics, China's critical industry information systems and information infrastructure have a demand for 700,000 talents of all kinds in the field of cyber security and the number is anticipated to increase to 1.4 million in 2020. However, by 2016, the number of those talents produced by higher education institutions was only less than 90,000. There are only about 10,000 graduates with bachelor, master and doctorate degrees in cyber space every year, so there is a big talent gap.[14] General Secretary Xi Jinping has pointed out, "We should input enough to invite excellent teachers, compile excellent textbooks, enroll excellent students and establish first-class cyber security academies." According to this instruction, China is speeding up the establishment of cyber security talents cultivation system.

1. **Polices on cyber security talents cultivation are being improved.**

Since 2005, the Chinese Government has launched a series of policies concerning cyber security talents cultivation, such as *Opinions on Further Enhancing the Construction of Information Security Discipline and Major as well as Relevant Talent Cultivation* and *Opinions on Promoting Informatization and Ensuring Information Security*. The Steering Committee for Information Security Teaching was established by the Ministry of Education in 2007 to enhance the disciplinary building concerning information security and to cultivate to a high-quality talent team in that respect. In June 2016, six ministries, namely, Office of the Central Leading Group for Cyberspace Affairs, National Development and Reform Commission,Ministry of Education, Ministry of Science and Technology, Ministry of Industry and Information Technology and Ministry of Human Resources and Social Security, jointly released *Opinions on Enhancing the Construction of Cyber Security Discipline and Relevant Talent Cultivation*, which proposes that the country should speed up the construction of the discipline and department of cyber security in higher education institutions, innovate the mechanism for cyber security talents cultivation, strengthen the compilation of relevant textbooks and the team-building of teachers in that field, promote collaborative innovation by higher education institutions and businesses, enhance on-the-job training of cyber security staff, and improve the public awareness and skills of cyber security and supporting measures for cyber security talents training. To sum up, talents guarantee will be provided for the implementation of the strategy on strengthening the country through cyberspace and safeguarding the national cyber security.

2. **The educational system concerning cyber security keeps being improved.**

Thanks to over one decade of development of the information security discipline, China now has its independent systems of basic theories, technologies and applications concerning cyber space education, and a number of secondary disciplinary research directions such as cryptology and its application, system security, network security and application system security. In June 2015, Academic Degrees Committee of the State Council and Ministry of Education jointly issued *Notice of Adding Cyber Security to the Primary Disciplines*, which proposed to add "Cyber Security" as a primary discipline to "Engineering", which means that graduates in "cyber security" would get the degree of engineering. In 2016, three majors,

[14]Source: CCID, *Blue Paper of China's Cyber Security Development (2016–2017).*

Table 6.2 New higher education institutions that can grant doctorate degree in cyber security as the primary discipline

No.	University	Remark	No.	University	Remark
1	Tsinghua University	new	16	Huazhong University of Science and Technology	new
2	Beijing Jiaotong University	new	17	Sun Yat-Sen University	new
3	Beijing University of Aeronautics and Astronautics	new	18	South China University of Technology	new
4	Beijing Institute of Technology	new	19	Sichuan University	new
5	Beijing University of Posts and Telecommunications	new	20	University of Electronic Science and Technology of China	new
6	Harbin Institute of Technology	new	21	Xi'an Jiaotong University	new
7	Shanghai Jiaotong University	new	22	Northwestern Polytechnical University	new
8	Nanjing University	new	23	Xidian University	new
9	Southeast University	new	24	University of Chinese Academy of Sciences	new
10	Nanjing University of Aeronautics and Astronautics	new	25	National University of Defense Technology	new
11	Nanjing University of Science and Technology	new	26	Information Engineering University	new
12	Zhejiang University	new	27	PLA University of Science and Technology	new
13	University of Science and Technology of China	new	28	PLA Electronic Engineering Institute	adjusted
14	Shandong University	new	29	Air Force Engineering University	adjusted
15	Wuhan University	new	–	–	–

Source Ministry of Education

namely, cyber security, information countermeasure and confidentiality management, had been opened at 121 points and majors related to network talents cultivation, such as electronic information and computer science, had been opened at over 4,800 points.[15] There are 29 universities that can grant doctorate degree in cyber security as the primary discipline (See Table 6.2), and 21 universities that

[15]Source: China Youth Daily, *China's First Special Fund for Cyber Security Has Been Set Up*, February 3, 2016.

have established school of cyber security.[16] In August 2017, Office of the Central Leading Group for Cyberspace Affairs and Ministry of Education jointly released *Measures for Demonstration Projects of First-class Cyber Security Academies' Construction*, proposing to launch the demonstration projects of first-class cyber security school construction from 2017 to 2027 to discover the new way, mechanism and system of cyber security talents cultivation and develop 4–6 cyber security schools acknowledged domestically and internationally. So far, the list of first seven universities with the first-class cyber security schools as demonstration projects has been published. They are Xidian University, Southeast University, Wuhan University, Beijing University of Aeronautics and Astronautics, Sichuan University, University of Science and Technology of China and Information Engineering University.[17]

3. **Varieties of cyber security contests are being launched.**

The cyber security contest is a means of discovering, cultivating and selecting cyber security talents and a key part in improving the cyber security education and training system. Through varieties of cyber security contests, the technical level of cyber security talents has been improved and a number of such talents have been discovered. First, governmental agencies have launched the information security and countermeasures contest. Since 2008, the Steering Committee of Information Security Teaching of Ministry of Education has launched 10 information security contests among university students to cultivate the students' awareness of cyber security innovation and team spirit and to popularize relevant knowledge among them. Since 2014, XCTF has been launched jointly by Cyber Security Talents Foundation and China Institute of Innovation and Development Strategy and organized by Nanjing Cyberpeace Information Technology Co., Ltd. The contestants are university and research institute students, technical staff from businesses and cyber security technology fans. Shanghai began to host ISG Information Security Skill Contest in 2009 and Hunan hosted 2016 Huxiang Cup Cyber Security Skill Contest. Secondly, universities are active in organizing countermeasures contests. For instance, Beijing Institute of Technology (BIT) has hosted the Information Security and Countermeasures Contest (ISCC) since 2014, and Xidian University has hosted XDCTF since 2009. Thirdly, Internet businesses or cyber security businesses organize similar contests. In February 2017, Tencent launched TCIF which divided into 0CTF and RisingStar CTF, covering areas such as crack, reverse, cryptology, web security, programming and mobile security. Alibaba has launched ALICTF, covering such areas as wireless terminal security,

[16]Source: Xinhua News Agency, *A Safety Dam Is Built for Online Communities: a review of five-year development of cyber security*, September 15, 2017.

[17]Source: Xinhua News Agency, *The List of First Group of Universities with the First-class Cyber Security Schools as Demonstration Projects Has Been Published*, September 11, 2017.

Linux kernel security, VPC network security, data protection security, web security, reverse engineering, security data analysis and code security.

4. **The special fund for cyber security has been established.**

In February 2016, China's first special fund for cyber security—Cyber Security Special Fund of China Internet Development Foundation had its donation ceremony held in Beijing, with its startup assets of RMB300 million *yuan* to be used to award talents, teachers, standards and textbooks in cyber security and to subsidize the study and life of excellent students in cyber security and talent cultivation in that field. By far, two sessions of Cyber Security Outstanding Talents Award, Cyber Security Talents Awards, Cyber Security Excellent Teachers Award and Cyber Security Scholarship have been launched and altogether one outstanding talent, 20 excellent talents and 18 excellent teachers have received award and 275 excellent undergraduates and 275 graduates have been subsidized.[18]

6.4 Cyber Security Protection Capacity and Awareness Have Been Dramatically Improved

Cyber security is for the people and by the people. Safeguarded by cyber security management and technology, China puts stress on critical information infrastructure protection and improvement of the whole society's cyber security awareness, so that all the people make effort to maintain cyber security.

6.4.1 The Critical Information Infrastructure Protection Capacity Is Being Enhanced

Critical information infrastructure in areas such as finance, energy, power, communication and transportation is the nerve center of economy and society as well as the key national strategic resource and the top priority of cyber security. *Cybersecurity Law of the People's Republic of China* defines the concept of "critical information infrastructure" and its scope. According to the law, "The State shall focus on protecting the critical information infrastructure in important industries and fields such as public communications and information services, energy, communication and transportation, water conservancy, finance, public services and e-government affairs and any other piece of critical information infrastructure whose damage, function loss or data leakage will result in serious damage of the national security and economy and the people's livelihood and public interests." In July 2017, Cyberspace Administration of China asked openly

[18]Source: China Internet Development Foundation.

Table 6.3 China's targets of cyber security protection in the past years

Time	Document	Description	Scope
September, 2004	*Implementation Opinions on Multi-Level Information Protection*	The focus was on the basic information network and key information systems concerning national security, economic pillars and social stability	Information systems of national affairs (office systems of CPC and governmental departments); information systems concerning national security and people's livelihood such as finance, banking, taxation, customs, auditing, industry and commerce, social security, energy, communication and transportation and national defense industry; information systems of education and research institutions; information systems in key information networks such as public communication and broadcast and TV transmission; key information systems of network administration centers and key websites and others
July 2012	*Opinions on Promoting Informatization and Ensuring Information Security*	information security in key areas	(1) Security of key information systems and basic information networks, including information systems in energy, communication and transportation, and finance, which are concerned with national security and people's livelihood; basic information networks including telecommunication networks, broadcast and TV networks and the Internet; wireless networks; Internet websites, addresses, domain names and connected organizations. (2) Governmental and classified information systems. (3) Industrial control systems. (4) Information resources and personal information
2013	*Notice of Carrying Out Critical National Information System Investigation*	critical national information systems	Level-4 information systems in the multi-level information security protection and big cross-provincial networks connecting the whole country in Level-3 information systems; information systems supporting the important systems providing large-scope services to the public (excluding governmental websites); and extremely important systems concerning national security, key economic sectors, social security and public interests

(continued)

Table 6.3 (continued)

Time	Document	Description	Scope
November 2016	*Cybersecurity Law of the People's Republic of China*	Critical information infrastructure	Infrastructure of public communication and information services, energy, communication and transportation, hydraulic engineering, finance, public services and e-government affairs; other information facilities whose damage, function loss or data leakage may result in serious damage of the national security, economy and people's livelihood and public interests
July 2017	*Regulation of Critical Information Infrastructure Security Protection (Consultation Draft)*	critical information infrastructure	The network facilities and information systems of the following organizations, whose damage, function loss or data leakage may result in serious damage of the national security, economy and people's livelihood and public interests: (1) Governmental agencies and organizations of energy, finance, communication and transportation, hydraulic engineering, health and medical care, education, social security, environmental protection and public utilities; (2) Networks including telecommunication, broadcast and TV and the Internet, and organizations providing cloud computing and big data, and other large public network information service organizations; (3) Research and manufacturing organizations in science and technology for national defense, large equipment, chemicals, foodstuffs and drugs; (4) Broadcast stations, TV stations and news agencies; (5) Other critical organizations

for opinions on *Regulation of Critical Information Infrastructure Security Protection*, which further defined the critical information infrastructure protection scope. At the same time, cyber security situation awareness and emergency response capacity has been improved. In June 2017, the Office of the Central Leading Group of Cyberspace Affairs modified *National Emergency Response Plan for Cyber Security Incidents*. Centering on the situation awareness of cyber security, incident analysis, and attribution and restoration capacity after an attack, the modification was aimed to improve the mechanism for cyber security monitoring and warning, emergency response and incident prevention and to perfect cross-industry and cross-department threat intelligence sharing mechanism. It encourages key industries and qualified provinces (municipalities directly under the Central Government and autonomous regions) to first establish the sharing mechanism and technically supporting platforms connected with national ones; to enhance cyber security emergency response by making emergency response plans, and to improve the capacity of incident tackling. China's targets of cyber security protection in the past years are shown in Table 6.3.

6.4.2 Cyber Security Awareness Keeps Improving

To facilitate the cyber security awareness and prevention and protection skills of the public, China, in recent years, have launched the "National Cyber Security Week" to create the atmosphere in which people can realize their responsibility for participating in cyber security protection. Since 2014, four sessions of the "National Cyber Security Week" have been launched by the Office of the Central Leading Group of Cyberspace Affairs, in collaboration with Public Relations Department, State Commission Office of Public Sectors Reform, Ministry of Education, Ministry of Industry and Information Technology, Ministry of Public Security, Ministry of Science and Technology, the People's Bank of China, State Administration of Press, Publication, Radio, Film and Television, China Association for Science and Technology, All-China Federation of Trade Unions and the Central Committee of the Communist Young League. Besides, a series of theme activities such as logo promotion, knowledge contests, experts interview and video collection have been launched centering on cyber security and online civilization, promoting the idea of "cyber security for the people and by the people". Over 100 million people have directly participated in those activities, while the activity information has covered 200 million people. According to a survey, 98% of Internet users regarded cyber security as "very important",[19] which shows that the awareness of cyber security keeps improving.

[19]Source: xinhuanet.com, A Survey about the Internet: To enhance self-protection awareness, how is your awareness of cyber security?.

Chapter 7
Active Participation in International Cyberspace Governance

Chinese Academy of Cyberspace Studies

The Internet has expanded the scope of national governance and created new space for people's life, making the world into a global village. On the other hand, there have occurred a series of problems, including the imbalance of development, unsoundness of rules and unreasonableness of order as well as information gap, intellectual property right infringement, cybercrime, network sniffing, network attack and cyberterrorism, which all pose challenges to national sovereignty, security and interest. President Xi Jinping attaches importance to international governance of cyber space, having illustrated his ideas of international governance of cyber space on many occasions, especially the "four principles" on promoting the global Internet governance system reform and "five proposals" on building a community of shared future in cyberspace to promote the construction of peaceful, secure, open and cooperative cyberspace and of a multilateral, democratic and transparent global Internet governance system.

7.1 China's Proposal on International Cyberspace Governance Has Been Universally Acknowledged

7.1.1 China Has More and More Ideas of the International Cyberspace Governance

China takes an active part in international cyberspace governance, committed to promoting all countries to develop and construct secure cyberspace together based on mutual respect and trust. In May 2014, at the first-stage conference of Summit on Interaction and Confidence-Building Measures in Asia, President Xi Jinping

Chinese Academy of Cyberspace Studies
Beijing, China

Chinese Academy of Cyberspace Studies (ed.), *China Internet Development Report 2017*,
https://doi.org/10.1007/978-3-662-57521-5_7

pointed out that security should be universal and that no country should seek for its own absolute security by sacrificing other countries' security. In July 2014, when delivering a speech at the Parliament of Brazil, he pointed out that China would like to work together with other countries to respect network sovereignty, safeguard cyber security, construct peaceful, secure, open and cooperative cyberspace and build a multilateral, democratic and transparent international Internet governance system. That was the first time a top leader of a country had explicitly proposed the principle on network sovereignty. In December 2015, in the keynote speech at the opening ceremony of the Second World Internet Conference, President Xi proposed "four principles" on promoting the global Internet governance system reform and "five proposals" on building a community of shared future in cyberspace, systematically and completely illustrating China's thoughts on Internet governance, which was a milestone in the history of the world Internet development. In November 2016, at the opening ceremony of the Third World Internet Conference, he further stressed in his video speech the necessary joint effort to build a community of shared future in cyberspace. He said, "China will work together with the international community to ensure the common well-being of humanity, uphold cyber sovereignty, promote more fair and equitable global Internet governance and bring about an open, inclusive and secure cyberspace that features equality, mutual respect, innovation and orderly development."

Column 24 Four Principles and Five Proposals

On December 16th, 2015, in the keynote speech made at the opening ceremony of the Second World Internet Conference, President Xi Jinping, after analyzing the features of today's world Internet development, proposed four principles and five proposals on the Internet development.

Four principles: Respect for cyber sovereignty. We should respect the right of individual countries to independently choose their own path of cyber development, model of cyber regulation and Internet public policies, and participate in international cyberspace governance on an equal footing. **Maintenance of peace and security**. Cyberspace should not become a battlefield for countries to wrestle with one another, still less should it become a hotbed for crimes. No double standards should be allowed in upholding cyber security. **Promotion of openness and cooperation**. We create more converging points of interests, growth areas for cooperation and new highlights for win-win outcomes. Efforts should be made to advance complementarity of strengths and common development of all countries in cyberspace so that more countries and people will ride on the fast train of the information age and share the benefits of Internet development. **Cultivation of good order**. Cyberspace must be governed, operated and used in accordance with law. In the meantime, greater efforts should be made to strengthen ethical standards and civilized behaviors in cyberspace. We should give full play to the role of moral teachings in guiding the use of the Internet.

Five proposals: First, speed up the building of global Internet infrastructure and promote inter-connectivity, to enable more developing countries and their people to share the development opportunities brought by the Internet. **Second, build an online platform for cultural exchange and mutual learning**. Together, we will promote the prosperity and development of cyber culture, which will enrich people's mind and thinking, and advance human civilization and progress. **Third, promote innovative development of cyber economy for common prosperity**. We will be able to spur the growth of worldwide investment and trade, and promote global development of digital economy. **Fourth, maintain cyber security and promote orderly development**. We should push for the formulation of international cyberspace rules accepted by all parties to fight cybercrimes and jointly uphold peace and security in cyberspace. **Fifth, build an Internet governance system to promote equity and justice**. International cyberspace governance should feature a multilateral approach with multi-party participation, so that the global Internet governance system reflects in a more balanced way the aspiration and interests of the majority of countries.

7.1.2 Strategic Documents on International Cyberspace Cooperation Have Been Released

International Strategy of Cooperation on Cyberspace ("the Strategy" for short hereinafter) was released in March 2017. With peace and development, and win-win cooperation as its theme and the building of a community of shared future in cyberspace as its goal, it is a complete illustration of China's thought on promoting international communication and cooperation in cyberspace. It advocates observing the purposes and principles of the *UN Charter* to ensure peace and security of cyberspace; adhering to sovereign equality and non-hegemony in cyberspace and non-interference in other countries' internal affairs; jointly formulating international rules of cyberspace and building a multilateral, democratic and transparent global Internet governance system; promoting advantage complementation and mutual development, and closing the "digital gap" and ensuring everyone's sharing of Internet development achievements. The Strategy proposes China's initiative in promoting and participating in the international cooperation in cyberspace from nine aspects, emphasizing that the country, while making itself into a strong country in cyberspace, will pursue the idea of new-type international relations with win-win cooperation as the core, and work with the international community in strengthening communication, deepening cooperation, constructing new partnership, and building a community of shared future for human kind. The strategy is a guiding document for China's participation in international communication and cooperation in cyberspace and a reference for all countries' cooperation in cyberspace.

7.2 China Is Constructing International Platforms for Internet Governance

China is playing a role of a big responsible country, active in building global and regional multi-level Internet governance platforms to introduce China's thoughts on Internet governance and push the governments, businesses and civil organizations of all countries, especially developing ones, to speak out their ideas, communicate with each other, build consensus, and deepen cooperation. It is contributing to the reform of the global Internet governance system and the building of a community of shared future in cyberspace.

7.2.1 China Has Initiated the World Internet Conference

Since its initiation in 2014, the World Internet Conference has become the most representative platform for international Internet governance, having witnessed fruitful achievements in idea exchange, technical exhibition, economic and trade cooperation and consensus. It is an international platform of interconnection between China and the world and a Chinese platform of international sharing and governance of the Internet. From 2014 to 2016, three sessions of the World Internet Conference attracted over 4,600 governmental officials, scholars, entrepreneurs and international organization leaders,[1] having produced great influence worldwide. In November 2014, the First World Internet Conference was held in Wuzhen, Zhejiang Province. President Xi Jinping sent his congratulation letter, in which he pointed out that the development of the Internet had brought about new challenges to a country's sovereignty, security and development interest, so it was imperative for the international community to face the challenges and seek joint governance of the Internet to achieve win-win results. In December 2015, he was present at the opening ceremony of the Second World Internet Conference and made a keynote speech, in which he pointed out "four principles" and "five proposals". At that conference, the High-Level Advisory Council of the World Internet Conference ("the High-level Advisory Council" hereinafter) was founded and Fadi Chehadé, former president and CEO of ICANN, and Ma Yun (Jack Ma) were elected co-chairmen of the Council. A group of leading figures of the industry, academic circles and research institutions of the Internet were selected as council members to give their opinions on the world Internet development. The organizing committee of the conference proposed *Wuzhen Initiative*, which had been discussed and passed by the High-level Advisory Council. The Initiative embodies the Internet spirit of "openness, equality, collaboration and sharing", which has attracted worldwide attention. During the Second World Internet Conference, the Light of Internet Expo

[1]Source: Data released from the press conferences of the three sessions of the World Internet Conference.

attracted nearly 260 businesses from the world to exhibit their achievements and over 80 special presentations were held. In November 2016, the Third World Internet Conference was held with the theme "Innovation-driven Internet Development for the Benefit of All—Building a Community of Common Future in Cyberspace". President Xi Jinping made a video speech at the opening ceremony. At the conference, the first Release Ceremony for World Leading Internet Scientific and Technological Achievements was held, covering AI, quantum computing and cyber security. All the displayed technologies were cutting-edge and most influential ones, forming a leading technological feast in the field of the Internet. The Third World Internet Conference also released *Wuzhen Report on World Internet Development 2016*, which summarized the achievement of the world Internet development and governance in one year, analyzed the situation the Internet development was faced with and the next focus of the world Internet development and governance. The Report was also aimed to push the Wuzhen Initiative forward.

7.2.2 China Is Active in Establishing Bilateral and Multilateral Dialogue Systems for International Cooperation in Cyberspace Governance

China is active in establishing bilateral and multilateral dialogue mechanisms concerning the Internet and in promoting its deep cooperation with other countries in cyberspace. It is trying to find new channels for international communication and set up flexible multi-dimensional dialogue systems for international cooperation in cyberspace governance.

1. **China keeps enhancing Sino-U.S. communications and dialogues in cyberspace.**

The communication and cooperation between China and the United States as two big countries in cyberspace not only benefit themselves, but also influence the reform and development of the international cyberspace governance system. In recent years, through a series of forums and conferences, China has established three-dimensional communication and cooperation mechanisms with the U.S. government, Internet businesses and think tanks. Since 2007, eight Sino-U.S. Internet Forums have been held. In September 2015, when meeting the representatives at the 8th Sino-U.S. Internet Forum, President Xi Jinping stressed that the two countries should carry out constructive dialogues and cultivate highlights of Sino-U.S. cooperation based on mutual respect and trust to make the cyberspace more beneficial to both peoples and the world. His remarks were well received by the entrepreneurs from both countries. Since September 2015, the two countries have organized high-level expert panel meetings on international rules of cyberspace. In April 2017, during the meeting of the leaders of two countries, four high-level dialogue mechanisms for law enforcement and cyber security were set up and the first dialogue was held in October 2017. The representatives from the two countries discussed how to combat cybercrime and reached consensus on that respect. Sino-US dialogue mechanisms concerning cyberspace have since been consolidated and developed.

2. **China keeps deepening cooperation in Internet governance with Russia.**

President Xi Jinping and President Putin signed in June 2016 the *Joint Statement on Cooperation in Information Space Development*, in which they point out that all countries, including China and Russia, have their common benefit and cooperation space in the field of Internet, so they should, on the basis of mutual respect and trust, carry out substantial dialogues and cooperation in ensuring cyber security and promoting cyberspace development. China and Russia have reached consensuses on respecting their respective cyberspace sovereignty, strengthening technological cooperation, countering cybercrime and cooperating in cyber security emergency response. This is of guiding significance to communication and cooperation in the field of the Internet between China and Russia and even between any other countries in the world. This kind of relationship in cyberspace is the model of new-type relationship between big countries in cyberspace cooperation.

3. **China is carrying out extensive Internet governance cooperation with Europe.**

In recent years, China and Europe have been deepening their cooperation in the field of Internet by establishing and improving their dialogue and communication mechanisms, such as China-EU Digital Economy and Cybersecurity Expert Working Group, China-EU Digital Cooperation Round Table, China-UK Internet Round Table and China-Germany Internet Industry Round Table. Digital economy, "Internet+", 5G technology and online children protection have become new highlights of China-Germany cooperation, playing an important role in promoting the cooperation in Internet between the two governments and industries.

4. **China is carrying out close cooperation with other developing countries in the field of Internet.**

China has successfully hosted China-ASEAN Information Harbor Forum to make China-ASEAN Information Harbor the guide and leader in the communication in terms of cyberspace between China and ASEAN; China-South Africa Internet Roundtable to deepen multi-level and multi-field cooperation in cyberspace between the two countries; and China-Arab States Expo Online Silk Road Forum to construct Ningxia Pivot Project on the China-Arab Online Silk Road to benefit the Chinese and Arabian peoples and to serve the Belt & Road Initiative.

7.3 China Is Strengthening Its Communication and Cooperation with International Organizations

China takes active part in international Internet governance and produces influence on its progress, playing an important role in the development and reform of global cyberspace governance platforms like the UN and WEF and of the Internet technology management institutions like ICANN.

7.3.1 China Attaches Great Importance to Cooperation with International Organizations

1. **China is active in Internet governance in the UN framework.**

The World Summit on the Information Society (WSIS) and Internet Governance Forum (IGF) are the most important dialogue platforms in the UN framework. The former is the important achievement of the UN in international cyberspace governance in the past 20 years. The Chinese Government and civil organizations participate in all the agendas of WSIS, producing influence on the global Internet governance progress. At the first-stage summit held in Geneva of Switzerland in December 2003, the Chinese delegate of 19 people were present, emphasizing the importance of the government in administering digital society and narrowing the digital gap. At the second summit held in the city of Tunis in November 2005, the Chinese representatives expressed the Chinese Government's ideas and proposals on building information society from four aspects, namely, promoting coordinated development, strengthening international cooperation, fully respecting the difference in social systems of different countries and following the cultural diversity in the world. In 2015, at the meeting reviewing the 10 years of WSIS achievements, the Chinese Government submitted three position papers. Besides, the Chinese Government and social organizations have participated in IGF forums of all sessions. Both WSIS and IGF launched their new ten-year initiative in 2016. At the G20 Hangzhou Summit held in the same year, participating countries reached the consensus that the Internet governance should follow WSIS achievements, emphasizing that governments, private sectors, civil society, technical institutions and international organizations should fully and actively get involved in International Internet governance in accordance with their roles and responsibilities.

2. **China is promoting ICANN internationalization.**

ICANN is in charge of the space allocation of IP addresses, designation of protocol IDs, management of gTLD, state and regional TLD systems and root server systems. China always attaches importance to the reasonable allocation of critical basic Internet resources, initiatively organizing ICANN meetings and promoting the ICANN management mechanism reform. It played a positive role in the transfer of America's management over IANA from administrative to judicial. CNNIC has become the only new generic top-level domain (New gTLD) data hosting service organization verified by ICANN in the Asia-Pacific Region, providing new gTLD registration service agencies and subscribers in China and the Asia-Pacific Region with better and localized service in data recovery, language communication and dispute arbitration. In June 2015, China participated in Yeti DNS Project released by ICANN at its 53rd conference and acted as the chief executive. The Project, based on totally new frameworks like IPv6, aims to break the number limit of existing international Internet root servers (13) and formulate better rules of next-generation Internet root server operation.

7.3.2 China Takes Active Part in the Work of International Organizations of Internet

1. **Chinese experts have important positions in international organizations.**

Hu Qiheng, an academician and President of Internet Society of China, was invited in 2004 by Kofi Atta Annan to be one of the 40 members of the UN Working Group on Internet Governance (WGIG). He contributed to the organization on behalf of China's Internet organizations. He was listed as the first Chinese into "Internet Hall of Fame" in 2013. Qian Hualin, Chief Scientist of Computer Network Information Center of Chinese Academy of Sciences, was elected ICANN Director in 2003, the first Chinese director, and he acted as Chair of APTLD. In 2014, he became the second Chinese listed into "Internet Hall of Fame". Zhao Houlin became Secretary General of International Telecommunication Union (ITU) in 2014, the first Chinese Secretary General of the organization in the past 150 years; Li Xing, Professor from Tsinghua University, had long been involved in the work of The Internet Engineering Task Force (IETF) before he was elected into IAB in 2013. Liu Dong, Director of Future Internet Engineering Center, acted as one of the directors of IEEE in 2015. Besides, a number of Chinese experts are in important positions of ICANN and IETF, contributing to the Internet governance of the world.

2. **Chinese businesses promote the progress of international affairs concerning Internet.**

Chinese businesses have joined all kinds of international organizations of Internet by giving full play to their strengths. Huawei is the representative of Chinese businesses involved in international Internet standardization. By December 2016, it had joined over 360 standard organizations/industrial alliances/open source communities taking over 300 important positions. It is a member of the board of directors of IEEE-SA, WFA, OneM2 M, OMA and OASIS and some other organizations. Baidu was the first Chinese business to have joined W3C, taking active part in the formulation of standards for network development platforms and wireless Internet. By December 2016, 32 Chinese businesses such as Qihoo 360, Tencent, Alibaba and China Mobile had become W3C members.[2]

[2]Source: updated data from http://www.chinaw3c.org/china-members.html.

Chapter 8
Benefits of the Internet to the People

Chinese Academy of Cyberspace Studies

According to General Secretary Xi Jinping, the development of cyberspace administration must be oriented to benefit the people. China always regards it as the starting point and foothold to make the Internet development serve the people, improve the people's welfare and promote the comprehensive development of human beings, aiming to speed up the popularization of informatization service, reduce the Internet application cost, alleviate poverty through the Internet, and provide the masses of people with available, affordable and good information service. The country makes full use of the characteristics of the Internet, such as inclusiveness, convenience and sharing, to facilitate resource sharing and improve public service so that the masses of the people will have more sense of gain in enjoying the benefits of the Internet development.

8.1 Popularization of the Internet Promotes Balanced Development Among Internet Users

After 23 years of development, the number of China's Internet users has reached 751 million, ranking first in the world, and the popularization rate of Internet has amounted to 54.3%. The Internet provides the opportunity of knowing about the world, obtaining information and realizing self-development for Internet users of different genders, ages, educational backgrounds, occupations, income and health conditions. The large number of Internet users, in return, lays a solid foundation for the sustaining development of China's Internet and becomes the inexhaustible force and power for its innovative development to benefit more people.

Chinese Academy of Cyberspace Studies
Beijing, China

Chinese Academy of Cyberspace Studies (ed.), *China Internet Development Report 2017*,
https://doi.org/10.1007/978-3-662-57521-5_8

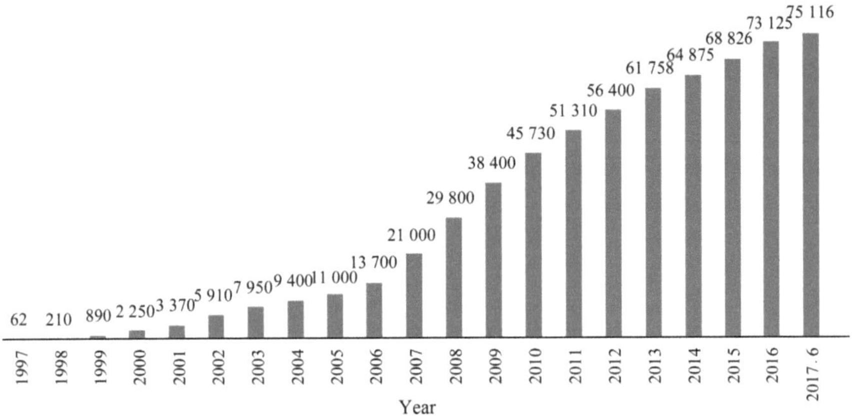

Fig. 8.1 Number of China's internet users (unit: 10,000) (1997-June 2017) (*Source* CNNIC)

8.1.1 The Number of Internet Users in China Keeps Increasing

1. **The number of China's Internet users is steadily growing.**

The number of China's Internet users has witnessed two growth periods: fast-growing period from 1994 to 2008 and plateau period (from 2009). In the first period, the number kept growing rapidly. In 1994, when China was fully connected into the Internet, only a small group of science and technology workers and research and technical staff could browse the Internet. By 2002, the number of Internet users in China had risen to 59.1 million, ranking second in the world, only next to that of the United States. In June 2005, the number was over 100 million and thus the Internet market scale was primarily formed and the industry began develop and innovate fast. In June 2008, the number reached 253 million, ranking first in the world. From 2009, it began to grow slowly, with the growth rate decreasing from 41.9% in 2008 to 28.9% in 2009. However, due to the large base, there was still an increase of 86 million. By June 2017, the number of China's Internet users was 751 million (See Fig. 8.1), the Internet popularization rate was 54.3% and the average Internet browsing time length was 26.5 h per week.[1] The Internet popularization rate of the world and China from 1997 to 2017 is shown in Fig. 8.2.

2. **The number of the mobile Internet users is increasing rapidly.**

With the popularization of the mobile Internet and the improvement of the quality of the content of mobile application programs, more and more Internet users have obviously begun to use the mobile Internet. Since 2009, as the mobile terminal

[1]Source: CNNIC.

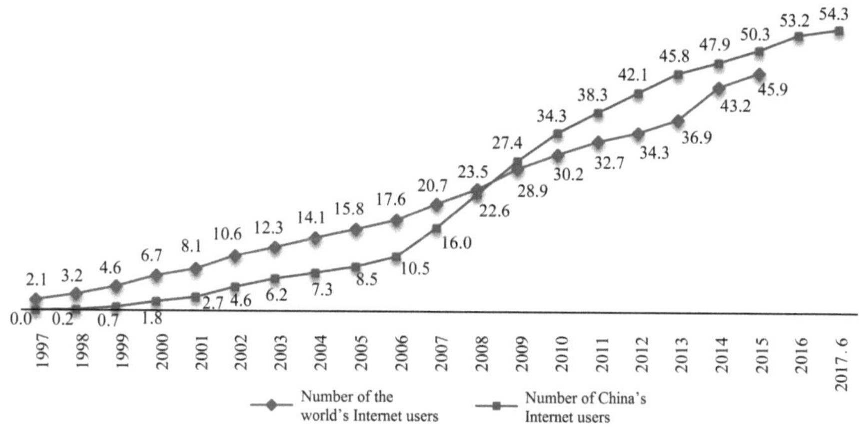

Fig. 8.2 Internet popularization rate of the world and China (1997-June 2017) (*Source* CNNIC and the World Bank)

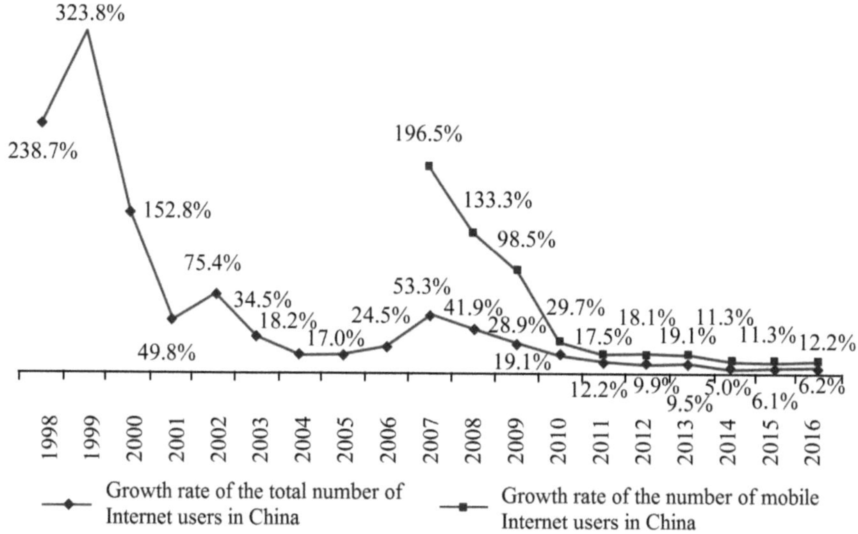

Fig. 8.3 Growth Rate of the number of mobile internet users and internet users in China (1997–2016) (*Source* CNNIC)

industry has seen its prosperity, the mobile communication network has seen its wide coverage, and especially the 3G network has increased the speed of mobile Internet surfing, the mobile Internet has been popular, leading to the steady growth of the number of the mobile Internet users. According to data released by Ministry of Industry and Information Technology, every 100 people in China had 0.3 mobile phones in 1995, but 98.7 ones in 2017. The popularization rate of terminal devices in first-tier cities has reached nearly one hundred percent and that in middle and

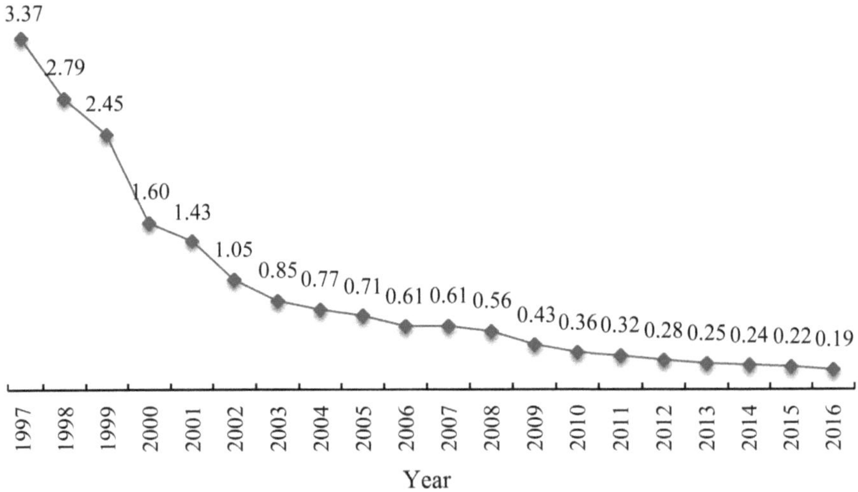

Fig. 8.4 Variation (coefficient of variation) between provinces in Internet popularization rate (1997–2016) (*Source* CNNIC)

western regions is growing dramatically. By June 2017, the number of mobile Internet users had amounted to 724 million, having seen an increase to 96.3%.[2] The growth rate of the number of mobile Internet users and Internet users in China from 1997 to 2016 is shown in Fig. 8.3.

8.1.2 Popularization of Internet Is Accelerating

1. **The gap in Internet popularization rate between provinces and between urban and rural areas is narrowing.**

From 1997 to 2016, the popularization rate of Internet in the whole country witnessed a dramatic increase. Through the variation coefficient analysis,[3] it can be seen that the difference between provinces in Internet popularization rate dropped from 3.37 to 0.19 (See Fig. 8.4) and the difference between regions also decreased. China attaches great importance to the popularization of Internet in rural areas, having worked out a series of reform and promotion measures to tackle the imbalance in the development of the Internet in early-liberated, minor ethnic group-inhabited, mountainous and poor areas. Every Village Project launched in

---

[2]Source: CNNIC.

[3]Coefficient of variation is defined as the ratio of the standard deviation to the mean. It shows the extent of variability in relation to the mean. The bigger the coefficient of variation of the provincial Internet popularization rate, the greater the difference in the Internet popularization rate between provinces.

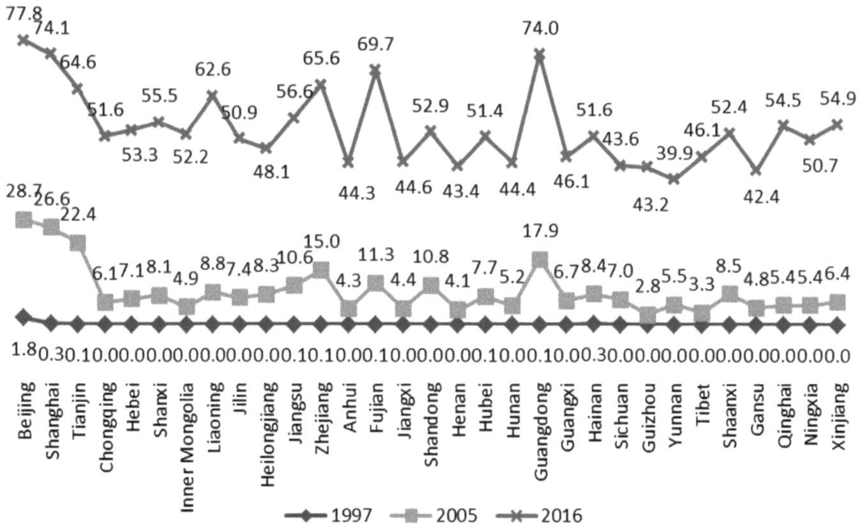

Fig. 8.5 Internet popularization rate (percentage) in all provinces (autonomous regions and municipalities) in 1997, 2005 and 2016 (*Source* CNNIC)

2004, Broadband Countryside Project launched in 2014 and the compensation mechanism for universal telecommunication service reform and innovation adopted in 2015 have all seen their success, leading to the leapfrog development of the information infrastructure in rural areas and the narrowing of the digital divide between urban and rural areas. According to data released at the press conference of industry and communication development held in October 2017, the total investment in two groups of pilot areas of universal telecommunication service by the end of the third quarter of 2017 had amounted to RMB40 billion *yuan*, covering the layout of fabric cables in 130,000 administrative villages, 43,000 of which are registered poverty-stricken villages. By the same time, the proportion of villages connected by broadband has exceeded 96% and the broadband coverage rate in poverty-stricken villages had reached 86%. The Internet popularization rate of all provinces (autonomous regions and municipalities directly under the Central Government) in 1997, 2005 and 2016 is shown in Fig. 8.5. The variation in the Internet popularization rate between urban and rural areas is shown in Fig. 8.6.

2. **The Internet contributes to targeted poverty alleviation.**

China has launched five projects concerning the Internet to promote targeted poverty reduction and alleviation, namely, Network Coverage, E-commerce in Rural Areas, Education Improvement through Internet, Information Service and Online Charity. The market plays an important role in promoting e-commerce in rural areas. There are comprehensive demonstration counties in terms of

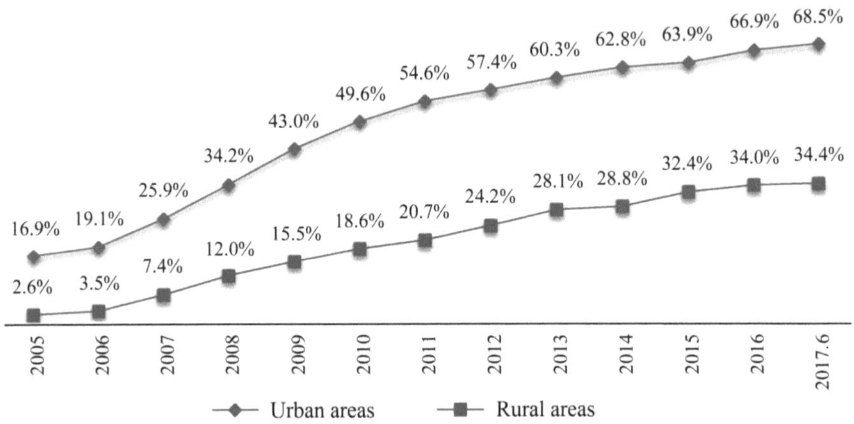

Fig. 8.6 Variation in Internet popularization rate between urban and rural areas of China (The population in urban and rural areas in June 2017 is an estimated number. It is the population in urban and rural areas at the end of 2016+ (the population in urban and rural areas at the end of 2016-the population in urban and rural areas at the end of 2015)/2.) (2005–June 2017) (*Source* CNNIC)

e-commerce's entry into rural areas and e-commerce has brought about an added income of RMB10 billion *yuan* for farmers.[4]

8.1.3 Progress Has Been Made in Information Accessibility Construction

In April 2008, *Law of the People's Republic of China on the Protection of Disabled Persons* was modified and passed. The Law stipulated that "The state and society shall take measures to gradually improve the barrier-free facilities, boost the barrier-free information communication, and create a barrier-free environment for disabled persons' equal participation in the social life." In July of the same year, *Technical Requirement for Barrier-free Information, and Barrier-free Technology of Website Design for Users with Physical Differences* formulated by Ministry of Industry and Information Technology were put into implementation. It was China's first industrial technical standard for information accessibility. Afterwards, a series of policies and regulations were launched, including *Guiding Opinions on Speeding Up the Construction of the Social Security System and Service System for the Disabled*, *Regulations on the Construction of Barrier-free Environments* and *Guiding Opinions on the Capacity-building of Barrier-free Website Services* were formulated and put into implementation to urge governments at all levels to launch information accessibility construction. Since 2005, China Disabled Persons

[4]*E-commercial development in rural areas nationwide from January to September*, 2017, http://scjss.mofcom.gov.cn/article/cx/201710/20171002660911.shtml.

Federation, China Foundation for Disabled Persons and Internet Association of China have organized the public service campaign that "100,000 blind people learn to use computers" and called on nearly 100 large domestic media websites to establish China Information Accessibility Alliance. At Beijing Olympics, Shanghai EXPO and Guangzhou Asian Games, Website Information Accessibility Initiative was launched. Businesses are active in innovating information accessibility technologies and products. In 2013, Tencent, Alibaba and Baidu, together with some other businesses, jointly set up China Information Accessibility Product Alliance to promote the development of information accessibility technologies and all kind of supporting software and new products welcomed by the disabled. China's Governmental Information Accessibility Popularization Initiative was launched in 2017. It is aimed to comprehensively complete the governmental information accessibility environment construction at the end of 2018, so that everybody can equally access to governmental information and service.

8.2 Internet Application Helps People to Enjoy More Benefits

8.2.1 Internet Medical Care Makes Health Service More Efficient

1. **Great progress has been made in medical informatization.**

Disease warning mechanisms and public health information systems have been continuously improved. In 2003, when SARS broke out, China Disease Control (CDC) system based on the Internet was set up, covering major disease control, such as infectious disease surveillance and public health emergency reporting. More than 98% of medical and health institutions above the county-level and 91% of township health centers can accurately monitor and report public health emergencies and emergency information, which is an important information support for decision-making, disease control and rapid response to emergencies. According to an investigation report on China's hospital informatization (2015–2016),[5] administration and clinical medical management are highly informatized in the country's hospitals. The implementation rate of HIS (Hospital Information System), such as the inpatient pharmacy management system, the pharmacy management system and outpatient and emergency registration system, has reached 70–75%; that of CIS (Clinical Information System), such as the resident nurse and doctor station information system and the electronic medical record (EMR) system, has reached 70%. The Internet has made medical and health service precise and personalized. Professional medical care institutions cooperate with Internet businesses in promoting online diagnosis, online health guidance and medical care resource sharing.

[5]The investigation covered 342 tertiary hospitals and 194 hospitals below.

All kinds of portable medical intelligent terminals and wearable devices make mobile medical care closer to common people.

2. **AI technology has primarily been applied to the medical area and big data is improving medical care efficiency.**

In recent years, Internet businesses have sped up the application of AI technology to the medical area and AI products have been gradually used in the clinical diagnosis and as the doctor's assistant. From 2016–2017, Baidu, Alibaba, Tencent and some other Internet businesses have launched their AI products into medical care, for instance, Baidu Medical Brain, ET Medical Brain and Tencent Medical Imaging. Through the collection, gathering, classification and sorting of a patient's symptoms, AI systems get massive medical data and simulate the clinical process through AI technical analysis and thus assist the doctors in completing the diagnosis, and they also act as assistants of doctors in medical imaging, new medicine release and health management. Meanwhile, traditional medical care institutions are trying to cooperate with Internet platforms to improve medical diagnosis efficiency by using big data technology. In 2017, the Second Xiangya Hospital of Central South University, DXY and Danale began their cooperation in AI-aided skin disease diagnosis. They have developed the AI-aided lupus erythematosus diagnosis model by collecting clinical data and historical data. Lupus erythematosus identification accuracy rate has reached over 85%.[6]

3. **Integrated online and offline development facilitates long-distance medical care.**

China's first Internet hospital was founded in Guangdong in 2014, marking the beginning of the Internet's reshaping the medical system. After Wuzhen Internet Hospital was founded at the end of 2015, Internet hospitals began to see the peak of their construction in China. More than 30 Internet hospitals were founded including the Internet Hospital of No. 1 Hospital afflicted to Zhejiang University and Guangdong Cloud Hospital. The online diagnosis and treatment platforms can provide not only online service but also offline service to increase the availability and interactivity of medical service. Since 2015, the online diagnosis and treatment platforms of Chunyuyisheng and DXY have established their offline clinics. In 2017, Tencent Doctorwork and "Medical Mall" in Hangzhou were established. Thanks to Internet hospitals and mobile medical platforms, long-distance consultation, diagnosis and monitoring and other medical services have been launched in China. According to data released by National Health and Family Planning Commission, by the end of 2016, totally over 6,800 hospitals had provided long-distance medical service, covering 1,330 counties. People now can share cross-hospital and cross-regional medical resources. Use efficiency of medical resources including doctor resources and medical devices has been improved. Since

[6]China's first AI-aided skin disease diagnosis system is released and "diagnosis through picture taking" becomes reality, http://www.sohu.com/a/141803141_114731.

there are problems like difficulty in registration, long waiting time for the diagnosis, short diagnosis time and payment difficulties, Internet medical service is now available in diagnosis, appointment registration and medicine purchasing to meet the users' demand. CNNIC data show that the use rate of online medical information inquiry, consulting for diagnosis and appointment registration accounts respective for 40.6, 39.1 and 24.1% of Internet medical service, which shows that medical care has become more convenient and satisfactory.

8.2.2 Internet Education Has Contributed to Equal Distribution of Educational Resources

1. **Three Accesses and Two Platforms have boosted rapid education informatization.**

In 2010, the State Council released the *National Planning Outline for Mid and Long-term Education Reform and Development (2010–2020)*, in which education informatization was elevated to a strategy of the nation. Through Three Accesses and Two Platforms (broadband network access to all schools, quality resource access to all classes and online learning access to everyone; the educational resource public service platform and the education management service platform), achievements have been made in China's education informatization. Data released by Ministry of Education show that from 2012 to 2016, the Internet access rate of China's primary and secondary schools rose from 25 to 87%, and the proportion of multimedia rooms rose from less than 40 to 80%, and the number of computers owned by every hundred primary and secondary school students rose from 8 to 12. Developed provinces and municipalities in the east, such as Jiangsu and Shanghai, have launched the wireless educational MAN (metropolitan area network), CN (campus network) and "future classroom" pilots, which have all contributed to the improvement of online teaching. The "the educational resource public service platform" was launched online at the end of 2012. By August 2017, totally 11.36 million teachers' rooms, 5.15 million students' rooms and 4.48 million parent's rooms had been opened on the platform, which had been connected with 22 provincial platforms and 23 city/county-level ones.[7] In 2016, in the Data Center of Ministry of Education, a complete structure of networks, storage, computing, security and operation and maintenance has been formed. Infrastructure construction has been completed in 32 provincial educational data centers, and the Ministry's Data Center and the provincial educational data centers are now interconnected.[8]

[7]*Monthly Report on Educational Informatization* (August 2017).

[8]*Progress of Educational Informatization* (2016), Ministry of Education.

2. **MOOC is thriving.**

As one of the early online education models, MOOC (Massive Open Online Courses) began to develop in China in 2013. Since then MOOC led by businesses and universities has been thriving. In the past four years, a number of influential online course platforms have been launched, such as MOOC of Chinese universities on icourse.cn, XuetangX of Tsinghua University, CNMOOC of Shanghai Jiaotong University and Chinese MOOCs of Peking University, with 1,400 MOOC courses and 5,600 SPOC courses, and over 30 million sign-up records and over two million credit-obtaining students for the online courses. In 2016, over 40 courses provided by top universities of China were opened on prestigious international platforms. At present, over 170 excellent MOOC courses of China are available in the world, and 992 high-quality video open classes and 2,886 high-quality resource-sharing classes are available online for free.[9]

3. **There are more varieties of online educational application products.**

There are more and more innovative services, such as live streams, recorded broadcasts and micro-lectures, as well as diverse online educational application products, such as online coaching, online question banks and online homework Q&A, which involve more and more students and facilitate teacher-student interaction. Big data, AI and other technologies also produce influence on online education. There have emerged the Uni Intelligent Learning System on hujiang.com, NCEE (National College Entrance Examination) Robot on xueba100.com, and AI English Teacher on liulishuo.com, which are applied in multiple educational scenarios. CNNIC data show that by June 2017, the number of online educational users in China had reached 144 million, with the half-a-year growth rate of 4.8%, and the online education use rate had reached 19.2%.

8.2.3 Internet Charity Boosts Innovation of Traditional Charity

In 2007, third-party online payment platforms like Alipay and TenPay extended their business into charity, changing the traditional model in which charity organizations could only rely on deposit through banks and donation, lowering the threshold for fund raising, and making it convenient for the public to donate money. Since 2013, thanks to WeChat and Sina Weibo and some other network media platforms, charity has been enjoying rapid spread and low cost. The Internet plays a positive role in money donation and charity concept spread, promoting China's charity to develop fast. In 2015, Tencent launched its Charity Hiker, an online step-donation project, in which the users can get eggs, milk and drinking water as

[9]Ministry of Education, *Progress of Educational Informatization* (2016).

donations after donating their recorded walking steps. By October 2017, 470 million Internet users had donated their steps in exchange for RMB580 million *yuan* of charitable fund.[10] In August 2016, Ant Forest of Alipay was launched to collect the "green energy" saved by the users by taking public transport and paying online so that Ant Financial plant sand-protecting trees like haloxylon ammondendron and salix mongolica in Alashan of Inner Mongolia on behalf of those low-carbon consumers. By the end of August of 2017, over 230 million users had reduced carbon emission by 1.22 million tons and 10.25 million trees had been planted, covering an area of 160,000 *mu*.[11] When earthquakes happened in Jiuzhaigou of Sichuan Province in August 2017, many institutions, including China Social Welfare Foundation (CSWF), launched first-aid projects on micro-platforms, raising over RMB three million *yuan* as donation.[12] In 2016, the charity platform of Tencent, Ant Financial and Taobao raised RMB 1.289 billion *yuan* as donation, with an increase of 37.79% in comparison with that in 2015.[13] Medical and education aid is the most attractive area in donation, then come disaster relief and reduction, and environmental protection. According to statistics, by October 2017, more than one billion donations had been done through the Internet, involving RMB two billion *yuan*, mostly by those born in 1980s and 1990s. 80% of the donation received by some charity institutions had been done through the Internet.[14]

Column 25 Vip Love Workshop (Weiai Gongfang)

Vip Love Workshop is a characteristic e-commerce charity platform launched by vip.com. By providing mature free-of-charge e-commercial platform support for charitable commodities, it can accelerate the marketization of those commodities and create the experience of "quality life starting from love" for its 300 million member users.

Vip.com inputs money in production at the early stage and provides free design, quality control, packaging, operation, promotion and logistics to support the charity platform, on which the charitable commodities can be sold with no profit for vip.com but all incomes are used to help the disadvantaged

---

[10]Source: Take the "Internet+" Express and Do Practical Charity, Qian Yibin, Zhao Dantong & Kuang Xixi, *People's Daily*, November 2nd, 2017.

[11]Source: Ali 95 Charity Week.

[12]Source: Increasingly Popular Charity through Finger Movement, Qian Yibin, Zhao Dantong & Kuang Xixi, *People's Daily*, November 2nd, 2017, http://paper.people.com.cn/rmrb/html/2017-11/02/nw.D110000renmrb_20171102_1-14.htm.

[13]Source: *Annual Report on China's Charitable Donation* (2016).

[14]Cnr.cn, Ministry of Internal Affairs: *Online donation is a new charity model, mostly done by those born in 1980s and 1990s*, Liu Huimin (journalist), October 15th, 2017.

groups of people like those poor mothers making the charitable commodities, so that they have a sustaining income maintaining their dignity.

Today, Vip Love Workshop has successfully launched online "Vip·Made by Mum—Tie-dyed", "Embroidered Fabric Made by Tujia Mum", and "Hometown Taste: Wheat from Black Earth". Both domestic and foreign designers are invited to redesign the platform from the perspective of the Eastern aesthetics to make traditional techniques more fashionable, practical and popular. The precious intangible cultural heritages like tie-dyed fabric made by mothers of the Bai Ethnic Group and the embroidered fabric made by mothers of the Tujia Ethnic Group are more and more recognized and favored by consumers.

Chapter 9
Prospect of China's Internet Development

Chinese Academy of Cyberspace Studies

At the new historic starting point, the Internet development of China should be guided by the Xi Jinping's Thought on Socialism with Chinese Characteristics for a New Era, the spirit of the 19th CPC Congress and the people-centered approach. We will speed up the implementation of the strategy of strengthening the country through the Internet, promote the construction of digital China, ensure the more complete and balanced, and fuller and better sustainable Internet development, so that the Internet can better benefit the country and its people.

9.1 Stronger Network Infrastructure

China will continue to implement the strategy of "Broadband China", accelerate the upgrade of information infrastructure, and promote fixed optical broadband construction, mobile broadband networking and next-generation Internet construction. It will speed up ultra-broadband optical access, 5G testing and commercialization, and IPv6 development while integrating the land and sea information Silk Roads and space-air-ground information networks, so that it can construct fast, mobile, safe and extensive advanced information infrastructure with integrated development of multiple networks and realize integration of multiple dimensions and comprehensive intelligent development. The country will continue to reduce the network expenses, improve the universal telecommunication service mechanism, speed up the construction of Internet infrastructure in rural areas, expand the coverage of optical fiber and broadband networks, increase Internet popularization and improve the people's quality in terms of information technology. Thus it can provide solid

Chinese Academy of Cyberspace Studies
Beijing, China

Chinese Academy of Cyberspace Studies (ed.), *China Internet Development Report 2017*,
https://doi.org/10.1007/978-3-662-57521-5_9

infrastructure for narrowing the digital divide, building an intelligent society and improving the people's sense of gain in sharing the Internet development achievements.

9.2 More Advanced Network Information Technologies

China will keep implementing the strategy of innovation-driven development and enhance basic and proactive researches for leading original breakthroughs in cutting-edge network information technologies. It will strengthen basic research in application, and implement key science & technology projects of network information while putting more manpower, things and funds in the R&D of core technologies. It will concentrate elite forces and make strategic arrangements to achieve breakthroughs in basic, general, asymmetric, cutting-edge and subversive technologies, so that it can catch up and become one of the leaders in core network information technologies.

9.3 More Prosperous Digital Economy

China will deepen the supply-side structural reform. Seizing the new opportunity of the integrated development of social informatization and economic globalization, it will facilitate the integration of Internet, big data and AI with real economy, so that the information flow will boost technical, capital, talent and material flows. It will also boost optimization of resource allocation, improvement of total factor productivity, and production, circulation and sales of commodities and service while reshaping the industrial structure and ecosystem, changing the economic growth mode and economic structure, activating innovative thinking of the whole society and providing new drive and energy for economic development. It will promote the innovative e-commerce development and cross-border e-commerce, so that all kinds of Internet-based innovation in industrial organization, business model, supply chain and logistic chain will emerge. The country will facilitate integrated development of software and hardware in the field of information, and encourage rapid development of new industries such as intelligent terminals, intelligent vehicles, smart devices, smart home, IoT, industrial Internet, online medical care and IoV. Cloud computing and big data will see further development. Leading cloud computing businesses will speed up the construction of their cloud ecosystems and enhance the dominance of cloud computing in all industries. Mobile cloud application will be popularized, mixed cloud technologies and products will be mature and industrial cloud platforms will develop dramatically. Leading businesses represented by Alibaba Cloud, Tencent Cloud and Huawei Cloud will keep launching their business overseas and their service export will represent the direction of China's service export. In the areas of big data, the national strategy of big data is being implemented and data resources openness and sharing are being accelerated. In China, the capacity is being improved of mining and analyzing big data through the comprehensive use of data resources to serve the economic and social development, the increase of social demand for big data will bring about the "golden time" of the industrial development, and the integrated innovation between big data technologies and other new-generation information technologies will become the dominant trend of the industrial development. The Internet is being integrated fast with manufacturing, production modes are becoming flexible, intelligent and precise, and intelligent manufacturing is facilitating the formation of new-type manufacturing systems. Information technology is being used more deeply in agriculture, which will facilitate the reform and marketing of agricultural production and circulation as well as the innovation of the value creation mode, and speed up the restructuring of the agricultural chain. Sharing economy is developing fast in China, with its industrial scale increasing and commercial mode being innovated. The Belt & Road Initiative will help to expand the space for international cooperation in digital economy. The construction of the platforms such as China-ASEAN Information Harbor, Online Silk Road Ningxia Pivot Project and China-EU Digital Silk Road will help Chinese businesses to speed up their "going out" into the overseas market.

9.4 Cleaner Cyberspace

Online and offline publicity will form concentric circles to guide and reflect public opinions. All forces of society and all ethnic groups will be united to make their effort to realize the Chinese Dream of the great rejuvenation of the Chinese nation. China will enhance the construction of the cyber content by launching relevant projects, promote the integrated development of traditional and emerging media to form a number of mainstream media and consolidate and strengthen online bases for culture and thought. Public opinions will be guided so that the cyberspace will be full of positive energy, including the core socialist values and other excellent Chinese cultural aspects. Healthy online culture will be introduced, capacity of online cultural innovation will be enhanced, and online cultural and artistic prosperity will be promoted. In this way, a number of online cultural brands with Chinese way and style will be launched. China will govern the cyberspace by laws. Relying on the improving Internet law systems, the country will enhance the cyberspace governance, clean the cyberspace, regulate the cyberspace order and counter illegal actions and activities contradictory to the people's interests, so that the Internet users, especially juveniles, can have clean and friendly cyberspace as their common home. The international transmission capacity of the Internet will be improved. Network media and social platforms will become important platforms for telling Chinese stories and transmitting the Chinese voice, thus contributing to the introduction of Chinese culture to overseas people. Both the government and businesses will collaborate with each other in cyberspace governance.

9.5 More Reliable Cyber Security

China will enhance top-level design and coordination in cyber security. In accordance with the *Cybersecurity Law of the People's Republic of China* and the *National Cyber Security Strategy*, the country will accelerate the improvement of the system of cybersecurity laws, regulations and policies, formulate and release the national strategy on credible cyberspace identity, and establish and improve the rules and regulations on critical information infrastructure protection, big data security management, cyber security censorship and personal data protection. It will make more effort to combat cybercrime and bad information spread to ensure the Internet users' legal rights and interests. China will deepen the administration of cyber security to monitor the cyber security all day long from all directions to enhance cyber security protection and deterrence. Cyber security inspection will be continued and unified and efficient mechanisms will be established for cyber security risk reporting, intelligence sharing, and judgment and handling, so that governments and businesses can share cyber security information. China's capacity of preventing cyber security risks will be improved through big data mining and analysis, so will its capacity of cyber security guarantee concerning critical information infrastructure and important information systems and personal information of citizens. The cyber security awareness and cyber literacy will be improved. It will become the common responsibility of the government, businesses, social organizations and Internet users to safeguard cyber security, so they will work together to make cyberspace cleaner and safer.

9.6 More Equitable International Governance Over Cyberspace

Cyberspace is the common space of human beings. All nations are connected with each other in that space, so their interests are also interconnected. It is in line with the interests of the international community and is the common responsibility of the international community to safeguard cyber peace and security and promote its openness and cooperation. Looking into the future, China will take "four principles" and "five proposals" put forward by General Secretary Xi Jinping as the guide and participate more in international communication and cooperation in cyberspace and promote the layout and management of root servers to be more internationalized and representative, and more in line with the world Internet development. The country will take its responsibility as a big county in terms of networks, and will contribute Chinese solutions and wisdom to the global cyberspace development and governance. It will give full play to the World Internet Conference, promoting the construction of new-type global digital economy platforms and cyberspace international governance platforms, and deepening multilateral, bilateral and multi-side communication and cooperation in cyberspace. It will contribute to building a multilateral, democratic and transparent global Internet governance system to ensure the Internet's more secure, full, balanced and sustainable development, so that it can better meet people's demand for a better life.

Looking into the future, China will continue to work for the goal of popularizing network infrastructure and enhancing independent innovation capacity. It will facilitate the comprehensive development of digital economy and the cyber security guarantee capacity, with broader mind, more inclusive philosophy, more innovative spirit and more practical measures. In this way, the country will achieve the great success of socialism with Chinese characteristics for the new era, deliver on its Two Centenary Goals and the goal of turning China into a modern strong socialist country and realize the Chinese Dream of great national rejuvenation, and thus contribute more to the well-being of mankind.

Appendix
Events of China Internet Development

In 2015, at the Second World Internet Conference, *China's Twenty-year Internet Development Report* was released, which gave a detailed account of Internet development events before 2012. Based on the *Report*, here we would like to summarize the events of China Internet development in the years after the 18th CPC National Congress.

2012

On January 18th, 2012, TD-LTE, proposed by Da Tang Telecommunications Technology & Industry Group, with China acting as the leader in its formulation, was defined by ITU as one of the international standards for the fourth-generation mobile communication.

On February 14th, 2012, Ministry of Industry and Information Technology released the *12th Five-Year Plan for the Development of Internet of Things*.

On March 27th, 2012, seven ministries including the National Development and Reform Commission jointly issued *Opinions on Next-generation Internet Development and Construction during the 12th Five-year Plan Period*.

In April 2012, people.cn went public on the A-share market, as the first listed key new central website, marking that China's network media construction entered a new period.

On May 9th, 2012, the Executive Meeting of the State Council reviewed and adopted the *Opinions on Promoting Informatization and Ensuring Information Security*.

On July 9th, 2012, the State Council issued the *Development Plan for National Strategic Emerging Industries during the 12th Five-year Plan Period*, in which "Broadband China" was proposed.

On September 18th, 2012, Ministry of Science and Technology released *12th-five-year Special Plan for Cloud Technology Development of the People's Republic of China* to accelerate the innovation and industrial development of cloud computing technology.

Chinese Academy of Cyberspace Studies (ed.), *China Internet Development Report 2017*,
https://doi.org/10.1007/978-3-662-57521-5

On September 18th, 2012, Internet Roundtable of Emerging Countries was held in Beijing, which was the first time when emerging countries had dialogues and communications on Internet.

On November 1st, 2012, organized by Internet Society of China, 12 search engine businesses signed the *Pact on Self-discipline of Internet Search Engine Service*.

On November 8th, 2012, the Report to the Eighteenth National Congress of the Communist Party of China pointed out that promoting harmonized development of industrialization, IT application, urbanization and agricultural modernization should be promoted, which was written into the *Constitution of the Communist Party of China*.

On December 5th, 2012, the First China-ROK Internet Roundtable was held in Beijing.

On December 28th, 2012, the Standing Committee of the National People's Congress reviewed and adopted the *Decision on Strengthening Information Protection on Networks*, requiring to protect personal e-information, prevent junk e-information and establish network identity management systems, and endowing competent authorities with necessary supervision power.

2013

On February 17th, 2013, the State Council released the *Guiding Opinions on Promoting the Orderly and Healthy Development of the Internet of Things*, and the National Development and Reform Commission and some other ministries jointly issued *Special Action Plan for Development of Internet of All Things* (2013–2015), in which 10 special action plans for development of Internet of All Things were launched.

In June 2013, the PRISM Scandal of the United States aroused the attention of the international community to information security and urged China to strengthen independent and controllable information security construction.

On June 25th, 2013, directed by Ministry of Public Security, 21 Internet businesses including Alibaba, Tencent, Shanda, NetEase and Amazon (China) jointly set up Internet Anti-fraud Committee to promote joint combat against online fraud and thus build a safe transaction ecosystem.

On June 26th, 2013, Internet Society announced the second list of Internet Hall of Fame, and Hu Qiheng (academician), President of Internet Society of China, was the first Chinese to be on the list.

On July 8th, 2013, China and the United States set up the Sino-U.S. Cyber Working Group in the framework of China-U.S. Strategic Security Dialogue, and held the first meeting in Washington.

On July 16th, 2013, Ministry of Industry and Information Technology released *Provisions on Protecting the Personal Information of Telecommunications and Internet Users*.

On August 1st, the State Council issued *"Broadband China" Strategy and Its Implementation Plan*, which proposes that broadband is the strategic public infrastructure of China and defines two-stage development goals of 2015 and 2020.

On August 14th, 2013, the State Council issued *Opinions on Promoting Information Consumption and Expanding Domestic Demand.*

On September 9th, 2013, the Supreme People's Court and the Supreme People's Procuratorate launched *Interpretation on Several Issues concerning the Specific Application of Law in the Handling of Defamation through Information Networks and Other Criminal Cases*, which was put into force on September 10th, 2013.

On October 25th, 2013, the *Law of the People's Republic of China on the Protection of Consumer Rights and Interests* (2013 Amendment) was released, stipulating that for the commodities sold through the Internet, TV, telephone or mail, the consumer has the right to return them within seven days upon receiving them, in order to enhance personal information protection and the responsibilities of online transaction platforms.

On November 9th, 2013, at the Third Plenary Session of the 18th Central Committee of the Communist Party of China, the *Decision of the CCCPC on Some Major Issues Concerning Comprehensively Deepening the Reform* was adopted to enhance cyber administration by law and speed up the improvement of the leading system for the Internet management.

On December 4th, 2013, China officially issued the first 4G licenses and China Mobile, China Telecom and China Unicom obtained their permit to operate 4G TD-LTE.

On December 26th, 2013, Ministry of Industry and Information Technology issued the first licenses for mobile communications resale pilots, and 11 private businesses obtained their virtual operation licenses.

In December 2013, the Second China-ROK Internet Roundtable was held in Seoul of ROK.

2014

On February 27th, 2014, Xi Jinping, General Secretary of CPC, President of the People's Republic of China, Chairman of Central Military Commission and Head of the Central Leading Group for Cyberspace Affairs, presided over the first meeting of the Central Leading Group for Cyberspace Affairs and delivered an important speech.

In April 2014, the National Office against Pornography and Illegal Publications, Cyber Administration of China, Ministry of Industry and Information Technology and Ministry of Public Security issued an announcement that from April to November of 2014, they would jointly launch nationwide the Special Campaign against Pornography and Illegal Publications: Cyber Cleaning 2014.

On June 12th, 2014, National Copyright Administration, Cyber Administration of China, Ministry of Industry and Information Technology and Ministry of Public Security launched Sword Net 2014, a special action to counter Internet piracy.

On July 16th, 2014, when delivering a speech at the Parliament of Brazil, president Xi Jinping pointed out that China would like to work together with other countries, on the principle of mutual respect and trust, to deepen international cooperation, respect network sovereignty, safeguard cyber security, construct peaceful, secure, open and cooperative cyberspace and build a multilateral, democratic and transparent international Internet governance system.

In July 2014, new Chinese generic top-level domains (GTLD) such as .公司 and 网络 were officially opened for registration.

On August 7th, 2014, Cyber Administration of China issued *Interim Provisions on the Administration of the Development of Public Information Services Provided through Instant Messaging Tools* to regulate the provision and use of instant messaging tools. It proposes explicit requirements on public information service with instant messaging tools.

On August 18th, 2014, the Fourth Meeting of the Central Leading Group for Comprehensively Deepening Reforms reviewed and adopted the *Guiding Opinions on Promoting the Integration of Traditional Media and Emerging Media.*

On October 21st, 2014, State Administration of Press, Publication, Radio, Film and Television and the Cyberspace Administration of China jointly issued the *Notice of the Verification and Issuance of Press Cards for News Websites* to strengthen the team building of news websites journalists.

On August 23rd, 2014, the Fourth Plenary Session of the 18th Central Committee of the Communist Party of China adopted the *Decision of the CPC Central Committee on Major Issues Pertaining to Comprehensively Promoting the Rule of Law*, which proposes to enhance legislation on the Internet, improve laws and regulations on cyber information service, security protection and social administration and regulate online actions in accordance with laws and regulations.

On November 6th, 2014, Cyber Administration of China and State Administration of Press, Publication, Radio, Film and Television launched the Special Campaign of Clearing up and Regulating Harmful Information in Online Video.

From November 19th to 21st, 2014, the First World Internet Conference jointly organized by Cyber Administration of China and Zhejiang Province was successfully held. President Xi Jinping sent his congratulations, Premier Li Keqiang met the participants and talked with them, and Vice Premier Ma Kai was present at the opening ceremony and made a keynote speech.

On November 24th, 2014, Office of the Central Leading Group for Cyberspace Affairs, together with some other ministries, held the Opening Ceremony for the National Cyber Security Publicity Week, at which Liu Yunshan, Standing Committee Member of the Central Political Bureau of CPC, Secretary of the Secretariat of the Central Committee of the CPC and Vice Head of the Central Leading Group for Cyberspace Affairs, was present and made an important speech. Ma Kai, Liu Qibao and Guo Shengkun were also present.

On April 8th, 2014, Qian Linhua, Researcher from Computer Network Information Center of Chinese Academy of Sciences, was listed into Internet Hall of Fame by Internet Society.

2015

On January 5th, 2015, the People's Bank of China issued the *Notice on Preparation of Individual Credit Collection*, marking the overall beginning of big data credit collection based on the Internet development.

On January 6th, 2015, the State Council issued *Opinions on Promoting the Innovative Development of Cloud Computing and Cultivating New Business Forms of the Information Industry*.

On February 4th, 2015, Cyber Administration of China released *Provisions on the Administration of Account Names of Internet Users*, i.e. *Ten Provisions on Account Names*.

On March 1st, 2015, China put into enforcement the *Pilot Program of Opening the Business of Broadband Access*, allowing private capital into broadband access. 16 cities including Taiyuan, Shenyang, Harbin, Shanghai, Nanjing, Hangzhou and Ningbo became the first pilot cities.

On March 2nd, 2015, the founding of the Chinese Culture Institute of Internet Communication was approved by Ministry of Internal Affairs.

In March 2015, "Internet +" was written into the R*eport on the Work of the Government*. Then the State Council issued *Guiding Opinions on Vigorously Advancing the "Internet Plus" Action*.

On April 15th, 2015, China's first big data exchange (Global Big Data Exchange) was founded in Guiyang, leading the exploration in data interconnection and sharing.

On April 28th, 2015, Cyber Administration of China launched *Provisions on the Interview of Entities Providing Internet News Information Services*, i.e. *Ten Provisions on the Interview*.

On April 29th, 2015, the Chinese Culture Institute of Internet Communication was founded.

On May 4th, 2015, the State Council issued *Opinions on Vigorous Development of E-Commerce to Accelerate Development of New Economic Driving Force*.

On May 8th, 2015, the State Council released *Made in China* (2025) to promote the implementation of the strategy of strengthening the country through manufacturing.

On August 31st, 2015, the State Council issued the *Action Outline for Promoting the Development of Big Data* to promote the development and application of big data in China and accelerate the making of China into a strong country of data.

On June 8th, 2015, Public Relations Department of Central Committee of CPC and Office of the Central Leading Group for Cyberspace Affairs held an advancing meeting, at which the participating websites and Internet businesses signed the *Proposal on Jointly Promoting Online Charity*.

On June 12th, 2015, Cyber Administration of China organized the first Network Integrity Publicity Day with the theme "Network Integrity with Me" to promote the idea of running and using the Internet with integrity.

On June 30th, 2015, Alibaba Chairman Ma Yun (Jack Ma) was elected Co-Chairman of Global Commission on Internet Governance.

On July 1st, 2015, *National Security Law of the People's Republic of China* was put into force, which proposes to enhance cyber administration, prevent, curb and punish by law cyber-attack, cyber-invasion, online confidential information theft and spread of harmful information to safeguard the cyber sovereignty, security and development interests.

On July 6th, 2015, the First China-EU Digital Cooperation Roundtable was held in Belgium.

On July 18th, 2015, 10 ministries, including the People's Bank of China, released *Guiding Opinions on Promoting the Sound Development of Internet Finance*.

On August 3rd, 2015, China Internet Development Foundation was founded in Beijing, marking the founding of China's (also the world's) first public-raising foundation in the field of Internet.

On September 13th, 2015, China-ASEAN Information Harbor Forum was held in Nanning, jointly organized by Cyber Administration of China, National Development and Reform Commission and Guangxi Zhuang Autonomous Region.

On September 23rd, 2015, the Eighth U.S.-China Internet Industry Forum was held in the United States. President Xi Jinping was present and delivered a speech.

On October 20th, 2015, the World Internet of Things Conference was held in Wuxi, Jiangsu Province.

On October 26th, 2015, the Fifth Plenary Session of the 18th Central Committee of the Communist Party of China adopted *Recommendations for the 13th Five-Year Plan for Economic and Social Development*, which proposes to implement the strategies of national security, strengthening of the country through the Internet, big data, and military and civilian integration, enhance the construction of online thought and culture communication, promote the integration between traditional and emerging media, and strengthen the capacity-building of international communication.

On November 1st, 2015, *Amendment (IX) to the Criminal Law* was promulgated, defining the obligations of network service providers for online information security management and intensifying the penalty on online information crime and the protection of personal information of citizens.

On November 6th, 2015, the first news website journalist permits were issued. 594 journalists from 14 central news websites including people.cn, xinhuanet.com and gmw.cn got their permits, becoming the first group of website journalists "with permits".

On November 23rd, 2015, the World Robot Conference was held in Beijing. President Xi Jinping sent congratulations on it.

On December 1st, 2015, the First U.S-China High-level Joint Dialogue on Cybercrime and Related Issues was held in Washington, co-chaired by Chinese State Councilor Guo Shengkun and U.S. Attorney General Loretta Lynch and Secretary of Homeland Security Jeh Johnson.

On December 16th, 2015, the Second World Internet Conference was held in Wuzhen, Zhejiang Province. President Xi Jinping was present and delivered a keynote speech, putting forward "four principles" on promoting the global Internet governance system reform and "five proposals" on building a community of shared future in cyberspace.

2016

On February 4th, 2016, State Administration of Press, Publication, Radio, Film and Television of the People's Republic of China and Ministry of Industry and Information Technology released *Provisions on the Administration of Online Publishing Services.*

On March 13th, 2016, Office of the Central Leading Group for Cyberspace Affairs launched the "To Be a Good Chinese Netizen" Project to call on Internet users to build clean cyberspace.

On March 25th, 2016, Cybersecurity Association of China was founded.

On April 19th, 2016, Xi Jinping, CPC General Secretary, President of the People's Republic of China, Chairman of Central Military Commission and Head of the Central Leading Group for Cyberspace Affairs, chaired a Symposium on Cybersecurity and IT Application and delivered an important speech. Li Keqiang and Liu Yunshan, Vice Heads of the Central Leading Group for Cyberspace Affairs, and Ma Kai, Wang Huning, Liu Qibao, Fan Changlong, Meng Jianzhu, Li Zhanshu, Yang Jiechi, and Zhou Xiaochuan were present.

On May 5th, 2016, Ministry of Ministry of Industry and Information Technology issued the Basic Telecommunication Business License to China Broadcast Network (CBN), who then can do business nationwide in domestic data Internet transmission and domestic communication facilities service.

On May 18th, 2016, National Development and Reform Commission, Ministry of Science and Technology, Ministry of Industry and Information Technology, and Office of the Central Leading Group for Cyberspace Affairs jointly formulated the *Implementation Plan for Three-year Action of "Internet Plus" Artificial Intelligence,* putting AI industry as a priority of development from the level of national strategy.

On June 14th, 2016, the Second U.S.-China High-level Joint Dialogue on Cybercrime and Related Issues was held in Beijing, and *the Operation Scheme of Hotline Mechanism for U.S.-China Combating Cybercrime and Related Issues* was adopted and the result list was jointly released.

On June 25th, 2016, China and Russia signed the *Joint Statement on Cooperation in Information Space Development*, seeking to promote information space development.

On June 25th, 2016, Cyberspace Administration of China released *Provisions on the Administration of Internet Information Search Services.*

On June 28th, 2016, Cyberspace Administration of China released *Provisions on the Administration of Mobile Internet Applications Information Services.*

On July 4th, 2016, State Administration of Industry and Commerce released *Interim Measures for the Administration of Internet-Based Advertising.*

On July 27th, 2016, General Office of the CPC Central Committee and General Office of the State Council released the *Strategic Plan for the National Informatization.*

From September 4th to September 5th, 2016, G20 Hangzhou Summit adopted *Digital Economy Development and Cooperation Initiative*, the world's first digital economy policy document signed by leaders of different countries.

On September 13th, 2016, according the first-instance judgement, Shenzhen Qvod Technology Co., Ltd. and the defendant were convicted of crime of spreading obscene articles for profit.

On November 1st, 2016, *Interim Measures for the Administration of Online Taxi Booking Business Operations and Services* was put into effect.

On November 4th, 2016, Cyberspace Administration of China released *Provisions on the Administration of Internet Live-Streaming Services.*

On November 7th, 2016, the 24th Session of the Standing Committee of the Twelfth National People's Congress of the People's Republic of China adopted *Cybersecurity Law of the People's Republic of China*, China's first basic law comprehensively regulating cyber security and a milestone in the country's cyberspace law construction.

From November 16th to November 18th, 2016, the Third World Internet Conference was held in Wuzhen, Zhejiang Province. President and CPC General Secretary Xi Jinping delivered an important speech through the video, and sent his congratulations. Liu Yunshan, Standing Committee Member of the Central Political Bureau of CPC, and Secretary of the Secretariat of the Central Committee of the CPC was present and delivered a speech.

On December 27th, 2016, Cyberspace Administration of China released *National Cyberspace Security Strategy*, China's first framework document on national cyber security.

2017

On January 10th, 2017, National Development and Reform Commission, Office of the Central Leading Group for Cyberspace Affairs and Ministry of Commerce and some other ministries issued *the Implementation Plan for Three-year Action of Promoting E-Commerce Development* (2016–2018).

On January 13th, 2017, the People's Bank of China released *Notice on Matters concerning Implementing the Centralized Deposit of the Funds of Pending Payments of Clients of Mobile Payment Institutions*, marking the establishment of the mechanism for centralized deposit of the funds of pending payments of clients of mobile payment institutions.

On January 15th, 2017, General Office of the CPC Central Committee and General Office of the State Council issued the *Opinions on Promoting the Sound and Orderly Development of the Mobile Internet*, defining the political directions and principles on promoting and regulating the development of mobile Internet.

On January 16th, 2017, nine ministries, including the National Development and Reform Commission and Ministry of Commerce, issued the *Guiding Opinions on Strengthening Honesty Construction in the E-commerce Field*, to strengthen the honesty construction in e-commerce and the measures guaranteeing the consumers' rights and interests.

On January 17th, 2017, Ministry of Industry and Information Technology issued the *Plan for the Development of Information Communication* (2016–2020).

On January 22nd, 2017, China Internet Investment Fund sponsored by Cyberspace Administration of China and Ministry of Finance was founded in Beijing, with the total expected fund of RMB 100 billion *yuan*, and the first raised fund of RMB 30 billion *yuan*.

On January 22nd, 2017, Ministry of Industry and Information Technology issued the *Notice on Cleaning Up and Regulating the Internet Access Service Market*, to regulate the market order, enhance network information management and promote the sound and orderly development of Internet industries.

On February 22nd, 2017, China Banking Regulatory Commission released the *Guidelines for the Online Lending Fund Depository Business*, to define the basic rules and implementation standards that online lending fund deposit should observe and encourage online lenders and commercial banks to carry out their business on the market principle of equality, voluntariness and mutual benefit.

On February 24th, 2017, China Merchants Bank announced that it had been successful in direct cross-border payment application through the block chain, marking the implementation of China's first project of block chain in the area of cross-border affairs, so it is a milestone in the domestic block chain finance. On May 16th, at the Block Chain Technology and Application Summit and China's First Block Chain Development Contest held in Hangzhou, China Block Chain Technology and Industry Forum announced the standard *Referential Framework of Block Chain and Distributed Ledger Technology*, which is the first basic block chain standard guided by the government.

On March 1st, 2017, Ministry of Foreign Affairs and Cyberspace Administration of China jointly released the *International Strategy of Cooperation on Cyberspace*.

On May 2nd, 2017, Cyberspace Administration of China released *Provisions for the Administration of Internet News Information Services*, making stipulations on Internet news information service licensing management, cyberspace information management systems, and responsibilities of Internet news information service providers.

On May 2nd, 2017, Cyberspace Administration of China released *Provisions on the Administrative Law Enforcement Procedures for Internet Information Content Management*, which was to come into force on and from June 1st, 2017.

On May 3rd, 2017, the world's first photon quantum computer surpassing early classic computers was launched in China. It had been developed jointly by *University of Science and Technology of* China, CAS-Alibaba Quantum Company Laboratory, Zhejiang University and Institute of Physics of China Academy of Sciences.

On May 14th, 2017, President Xi Jinping was present at the Opening Ceremony of the Belt and Road Forum for International Cooperation, and delivered a keynote speech entitled *Work Together to Build the Silk Road Economic Belt and The 21st Century Maritime Silk Road*. He pointed out, "We should pursue innovation-driven development and intensify cooperation in frontier areas such as digital economy, artificial intelligence, nanotechnology and quantum computing, and advance the development of big data, cloud computing and smart cities so as to turn them into a digital silk road of the 21st century."

On May 22nd, 2017, Cyberspace Administration of China released the *Rules for the Licensed Management of Internet News Information Services,* which was put into force on and from June 1st, 2017.

On June 1st, 2017, the *Cybersecurity Law of the People's Republic of China* was officially put into force.

On June 26th, 2017, the 36th Meeting of the Central Leading Group for Comprehensively Deepening Reforms reviewed and adopted the *Plan for Hangzhou Court of the Internet*. The court was established on August 18th, 2017 as the first court of the Internet in China.

On June 28th, 2017, China Banking Regulatory Commission, Ministry of Education, Ministry of Human Resources and Social Security jointly issued *Notice on Enhancing the Standardized Management of Campus Lending*, requiring that "all campus lending by online lending institutions be suspended and the lending already done be settled properly."

On July 3rd, 2017, eight ministries including the National Development and Reform Commission jointly issued the *Guiding Opinions on Promoting the Development of the Sharing Economy.*

On July 8th, 2017, the State Council issued the *Development Plan on the New Generation of Artificial Intelligence*, which marked the first time AI development was taken as a national strategy.

On August 25th, 2017, the Cyberspace Administration of China released *Provisions on the Administration of Internet Forum and Community Services* and *Provisions on the Administration of Internet Comments Posting Services*, both of which were put into force on and from October 1st, 2017.

On September 4th, 2017, seven ministries and institutions including the People's Bank of China and Cyberspace Administration of China jointly announced that from that very day, all kinds of fund raising in token money should be stopped, that those organizations having raised fund in that way should return all the money to protect the investors' rights and interests and tackle the risks, and that any fund raising through token money not stopped and any breach of laws or regulations in completed fund-raising projects with token money upon the announcement would be punished by law.

On September 7th, 2017, Cyberspace Administration of China released *Provisions on the Administration of Internet User Public Account Information Services* and *Rules to Regulate Internet Group Information Services*, both of which were put into force on and from October 8th, 2017.

On September 18th, 2017, Wu Jianping, Academician of Chinese Academy of Engineering, was listed by Internet Society into Internet Hall of Fame.

On September 29th, 2017, Beijing-Shanghai Trunk, the world's first quantum confidential communication network, was officially put into use.

Afterword

The 19th National Congress of the Communist Party of China put forward the Two-Step Development Strategy, ushering in the new era of socialism with Chinese characteristics. Especially, the Congress made an important strategic layout for promoting the development of the Internet, big data, AI and sharing economy and for making China into a strong country in cyberspace, a digital country and an intelligent society, providing fundamental principles and guidelines for the development of the Internet. In the new era, we have the new march and new mission. It is our hope that *China Internet Development Report* 2017 (the "*Report*" for short hereinafter) is an interpretation of Xi Jinping's thoughts on socialism with Chinese characteristics, especially his strategic thoughts of strengthening and developing the country through the Internet, an exhibition of China's achievements in Internet development, a systematic summary of Chinese experience in Internet development and governance, and a scientific outlook on China's Internet development prospect, so that it can better boost the Internet development. We also hope that the *Report*, through a detailed study of China Internet, will contribute Chinese experience and wisdom to the world Internet development and governance.

The compilation of the Report has won support and guidance from Cyberspace Administration of China, especially guidance from the leaders of the Administration and support in terms of data and material from the departments and institutions of the Administration. China Academy of Cyberspace, as the coordinator, has invited a number of think tank institutions to participate in the compilation, including China Network Information Center, Zhejiang Institute of Industry and Information Technology, China Academy of Information and Communication Technology (CAICT), the First Research Institution of Ministry of Industry and Information Technology, Security Center of Ministry of Industry and Information Technology, State Information Center, CCID, Institute of National Governance of Tsinghua University, Internet Development Research Center of Peking University, and Cyberspace Governance Research Center and Internet Laboratory of Fudan University.

The publication of the *Report* is also owed to the support and help from society. Due to our limited experience, capability and time for the compilation, there are probably errors in it. We sincerely hope that governmental agencies, international organizations, research institutions, Internet businesses, social associations, and people from all walks of life will offer their opinions and suggestions and provide more detailed materials so that we can make modifications and improvement to better support China Internet development.

China Academy of Cyberspace
December 2017

MIX
Papier aus verantwortungsvollen Quellen
Paper from responsible sources
FSC® C105338

If you have any concerns about our products,
you can contact us on
ProductSafety@springernature.com

In case Publisher is established outside the EU,
the EU authorized representative is:
Springer Nature Customer Service Center GmbH
Europaplatz 3, 69115 Heidelberg, Germany

Printed by Libri Plureos GmbH
in Hamburg, Germany